T0269541

# A Study of Black Hole Attack Solutions

# A Study of Black Hole Attack Solutions

## On AODV Routing Protocol in MANET

Elahe Fazeldehkordi

Iraj Sadegh Amiri

Oluwatobi Ayodeji Akanbi

University of Malaya, Kuala Lumpur, Malaysia

Matthew Neely, Technical Editor

AMSTERDAM • BOSTON • HEIDELBERG • LONDON
NEW YORK • OXFORD • PARIS • SAN DIEGO
SAN FRANCISCO • SINGAPORE • SYDNEY • TOKYO

Syngress is an imprint of Elsevier

SYNGRESS.

Syngress is an imprint of Elsevier
225 Wyman Street, Waltham, MA 02451, USA



ISBN: 978-0-12-805367-6

**British Library Cataloguing-in-Publication Data**
A catalogue record for this book is available from the British Library

**Library of Congress Cataloging-in-Publication Data**
A catalog record for this book is available from the Library of Congress

For Information on all Syngress publications
visit our website at store.elsevier.com/Syngress

# CONTENTS

# LIST OF TABLES

# LIST OF FIGURES

# PREFACE

Mobile wireless ad hoc networks (MANETs) are non-concentrated wireless networks that can be formulated without the need for any preexisting infrastructure in which every node can behave as a router. For transferring information to nodes that are out of its transmission range, the node should detect its local neighbors and through them, it will do transmission. MANET's characteristics like open medium, dynamic topology, lack of clear lines of defence makes the network so vulnerable to various attacks. One of the best and popular routing protocols that is used in MANET is ad hoc on-demand distance vector routing (AODV). AODV is gravely influenced by a famous attack in which a malicious node sends a forged route reply message that it has a short and fresh route to the destination node; this attack is known as black hole attack. In this book, MANET performance against both single and multiple black hole attacks has been investigated. And then intrusion detection system AODV routing protocol (IDSAODV) that is presented by Dokurer at 2006, has been implemented on the network. The result are analysed through different network parameters: total drop packets, end-to-end delay, packet delivery ratio, and routing request overhead, using NS-2.35. The results demonstrate IDSAODV solution method which was presented for the single black hole attack before also can be used effectively for reducing total drop packets and improving packet delivery ratio versus multiple black hole attacks. But, this solution method does not have considerable effect to improving end-to-end delay and routing request overhead. This testing only used UDP packets and did not test this protection when TCP packets are in use.

CHAPTER 1

# Introduction

## 1.1 INTRODUCTION

Mobile ad-hoc networks (MANET) are independent and noncentralized wireless techniques. MANETs involve mobile nodes which are free in shifting in and out in the network. Nodes are the techniques or gadgets, that is, cell phone, laptop computer, individual electronic devices, MP3 player, and PC that are forming the network and are mobile. These nodes can work like host/router or both at the similar time. They can form irrelevant topologies based on their connection with each other in the network. These nodes have the capability to set them up, and because of their self-settings capability, they can be implemented quickly without the need for any infrastructure (Ullah & Rehman, 2010).

Security in MANET is the most serious issue impacting performance of network. The accessibility to network services, protecting privacy, and ensuring reliability can be carried out by guaranteeing which protection problems have been met (Ullah & Rehman, 2010). MANETs generally are difficult to protect from attacks because of its functions like open medium, modifying its topology dynamically, deficiency of main monitoring and control, cooperative methods, and no obvious protection procedure (Ullah & Rehman, 2010). These aspects have modified the attack areas for the MANETs versus the protection risks.

Routing protocols have been created that determine how routers communicate with each other and how to select routes between any two nodes on a computer network. In general, routing methods is one of the complicated and exciting analysis places. Many routing methods have been designed for MANETS, that is, AODV, OLSR, DSR, etc. (Ullah & Rehman, 2010).

AODV is one of the well-known On-Demand Routing techniques (Das, Belding-Royer, & Perkins, 2003). Some scientists (Deng, Li, & Agrawal, 2002; Ramaswamy, Fu, Sreekantaradhya, Dixon, & Nygard, 2003) investigated this routing protocol and discussed the weaknesses in ad hoc routing protocols and the attacks which can be performed. According to research carried out by Usha and Bose (2012), AODV technique is the most unprotected against the black hole attacks.

## 1.2 PROBLEM BACKGROUND

MANET is well known due to the point which these networks are powerful, infrastructure less, and scalable. Despite the truth of reputation of MANET, these networks are so much revealed to attacks (Lu, Li, Lam, & Jia, 2009; Ullah & Rehman, 2010). Wi-fi hyperlinks also create the MANET more vulnerable to attacks which make it simpler for the enemy to go within the network and capture accessibility the continuous interaction. Different types of attacks have been examined in MANET and their impact on the network. Attack such as greyish opening, where the enemy node acts maliciously for enough time until the performance is decreased and then change to their regular actions. MANET's routing methods are also being utilized by the assailants by means of surging attack, which is done by the enemy either by using route request (RREQ) or details surging (Ullah & Rehman, 2010).

In any networks, the sender wants his/her information to be sent as soon as possible in a protected and quick way. In a wormhole attack, assailants promote themselves to have the quickest and greatest data transfer rate available, and the enemy places themselves in powerful place in the network. They create the use of their place, that is, they have quickest direction between the nodes (Mahajan, Natu, & Sethi, 2008; Shanthi, Lganesan, & Ramar, 2009). One of the upcoming problems in MANET is the restricted power supply; assailants take advantage of this weakness and attempts to keep the nodes online until all its power is depleted and the node go into long-lasting rest. Many other attacks in MANET, for instance, jellyfish attack, modification attack, misrouting attack, and routing table overflow have been analyzed and revealed (Ullah & Rehman, 2010).

In black hole attack, a harmful node uses its routing technique to be able to promote itself for having the quickest direction to the place node or to the bundle it wants to identify. Furthermore, this aggressive node promotes its accessibility to clean tracks regardless of verifying its routing table. In this way enemy node regularly will have the provision in responding to the direction demand and thus identify the details bundle and maintain it (Ullah & Rehman, 2010).

Researchers have suggested alternatives to recognize and remove black hole nodes (Deng et al., 2002; Ramaswamy et al., 2003). Deng et al. (2002) analyses in detail one type of attack called the "black hole" problem that can easily be employed against the MANETs. But they have not addressed the supportive black hole attacks. As stated in their method, details about the next hop to place should be involved in the route reply bundle when any advanced node responses for RREQ. Then, the resource node delivers a further demand (FREQ) to next hop of responded node and requests about the responded node and direction to the place. By using this technique, we can recognize existence of the responded node only if the next hop is reliable. However, this remedy cannot avoid supportive black hole attacks on MANETs. For instance, if the next hop cooperates with the responded node, as well, the response for the FREQ will be basically "yes" for both concerns. Then the resource will go forward and deliver details through the responded node which is a black hole node.

Ramaswamy et al. (2003) suggested a remedy to protecting versus supportive black hole attacks. Also, they claimed that no models or performance assessments have been done. Therefore, this project focuses on assessment of the performance of the suggested plan to protecting versus supportive black hole attack.

## 1.3 PROBLEM STATEMENT

Based on researches carried out, Sharma and Gupta (2009) shows that AODV greatly suffers from multiple black holes in terms of packet delivery ratio, drop packets, average end-to-end delay, and route request overhead. Besides, the most common techniques used are inefficient in responding to multiple black hole attacks and just can prevent from single black hole attack (Deng et al., 2002; Lee, Han, & Shin, 2002; Sun, Guan, Chen, & Pooch, 2003). Therefore, little attention has been given to examine and implement the existing methods for prevention of multiple black hole attacks. There is a need to analyze these methods on multiple black hole attacks. Therefore, this study will address the following questions:

1. How to detect single and multiple black hole attacks?
2. How to mitigate a single black hole attack using the most efficient solution?
3. How to mitigate multiple black hole attacks using the methods in Tsujii and Itoh (1989)?
4. How to determine the efficiency of the solution used in black hole attack by comparing IDSAODV with black hole AODV using the following metrics: packet delivery ratio, packet loss percentage, average end-to-end delay, and route request overhead?

## 1.4 INTENT OF STUDY

In this research, performance of one of the most efficient solutions for preventing single black hole attack in MANET using AODV routing protocol will be investigated in terms of packet delivery ratio, packet loss percentage, average end-to-end delay, and route request overhead. Then, we will examine MANET performance under multiple black hole attacks with proposed solution. At the end of this investigation, it will be highlighted if the solution which shows good performance in terms of high packet delivery ratio, low packet loss percentage, low average end-to-end delay, and reduce route request overhead for single black hole attack can also be a useful face with multiple black hole attacks.

## 1.5 AIMS

There are four objectives for this project. They are:

1. To investigate the existing solutions for preventing single black hole attack in MANET using AODV routing protocol.
2. To determine one of the efficient existing solutions above using four metrics: packet delivery ratio, packet loss percentage, average end-to-end delay, and route request overhead.
3. To implement the existed solution (IDSAODV (Dokurer, 2006) which has been presented for single black hole attack before) for multiple black hole attacks.
4. To compare the effects of above solution on MANET performance with single black hole attack and multiple black hole attacks.

## 1.6 SCOPE

The scopes of this research are as follow:

1. The project will study the effects of multiple black hole attacks in MANET using (AODV ad-hoc on demand distance vector) routing protocol.
2. Analysis of solutions for preventing single black hole attack is taken into account.
3. The impact of solutions for preventing multiple black hole attacks on the performance of MANET is evaluated, finding out if solutions for single black hole attack are also useful for multiple black hole attacks.
4. Simulation will be done in NS-2.35 (network simulator).
5. Simulation will examine on MANET without black hole attack, MANET with 1 black hole node, MANET with 2 black hole nodes, and MANET with 3 black hole nodes. And if each scenario will work on networks with 6 nodes, 20 nodes, and 30 nodes separately.
6. Simulation parameters will be obtained from authors of proposed solution (Ahmad, Manan, & Jalil, 2012; Arya & Jain, 2011).
7. The measurements will be obtained using packet delivery ratio, packet loss percentage, average end-to-end delay, and route request overhead.

## 1.7 THE SIGNIFICANCE OF THE BOOK

Nowadays, there is an increasing need to preventing multiple black hole attacks due to the adverse effect they can have on their victims. Lots of work has been done on single black hole prevention using several techniques to achieve the same goal, but the question is whether these techniques can be useful for multiple black hole attacks? This study evaluates the performance of one of the most efficient of these techniques as regards to packet delivery ratio, packet loss percentage, average end-to-end delay, and route request overhead for preventing of single black hole attack and multiple black hole attacks on MANET using AODV routing protocol by studying each of those metrics individually and will carry out by MANET performance with multiple black hole attacks under single black hole attack prevention solutions.

## 1.8 ORGANIZATION OF THE BOOK

The book consists of six chapters. Chapter 1 describes the introduction, background of the study, research objectives and questions, the scope of the study, and its primary objectives. Chapter 2 reviews available and related literature on black hole attack detection. Chapter 3 describes the study methodology along with the appropriate framework for the study. Chapter 4 describes the effects of the single black hole and multiple black hole attacks on MANET performance. Chapter 5 discusses the implementation, result, and analysis based on research framework. Finally, Chapter 6 concludes the book with a closing remark, recap of objectives, contribution, and future work.

CHAPTER 2

# Literature Review

## 2.1 INTRODUCTION

This chapter primarily reviews the available literature in the field under study. Accordingly, it will account for the definitions of concepts and issues that affect preventing cooperative black hole attacks in mobile ad hoc networks (MANETs) with ad hoc on demand distance vector routing protocol (AODV) routing protocol. The first part of this chapter will delve on describing wireless networks, ad hoc networks, and MANETs. The second part of this chapter will present different protocols in MANET and will more explain AODV routing protocol. The third part of this chapter will deal with existing vulnerabilities and attacks that are related to AODV routing protocols in MANETs and will explain in detail about black hole attack. The fourth part and the final part of this chapter will review earlier works related to black hole attack detection in MANETs using AODV routing protocol.

## 2.2 NETWORK

Before going into the details of wireless network, it is important to know about a network and different kind of networks attainable today. Network is any collection of devices/computers connected with each other by means of transfer channels that help the users to share resources and transfer information with other users. There are two main types of networks today, that is, wired networks and wireless networks.

## 2.3 WIRED NETWORKS

Those networks in which computer devices attached with each with the help of wire are wired networks. The wire is used as intermediate of communication for transferring the information from one spot of the network to other spot of the network.

## 2.4 WHY WIRELESS NETWORKS?

Wireless networks are getting more and more common because of their comfort of use. Consumer/user is no more rely on cables, so it is so easy to move from one place to another and enjoy being connected to the network. One of the great characteristics of wireless networks that make them attractive and different between the traditional wired networks is movability. With this characteristic the consumer has the ability to move without limits, while connecting to the network. Wireless networks are relatively easy to install compared with wired network. While using wireless networks, there is no need to worry about pulling the cables/wires in wall and ceilings. Wireless networks can be set up according to the need of the consumers. These can extend from small number of consumers to large networks that the number of users is in thousands. Wireless networks are very useful especially for areas where the wire cannot be installed like hilly areas.

## 2.5 WIRELESS NETWORKS

Nowadays, wireless networks are achieving popularity to its peak, because users want connectivity irrespective of their geographic location. With wireless networks, users are able to communicate and transfer data with each other without any wired intermediary between them. Greatly insinuation of wireless devices is one of the reasons of popularity of these networks. Wireless applications and devices principally emphasize on wireless local area networks (WLANs). This has essentially two manners of performances, that is, in the attendance of control module (CM) also known as base station and there is no CM in Ad hoc connectivity. Because of carrying out their performances, ad hoc networks do not rely on fixed infrastructure. In these networks the performance mode is stand alone or perhaps attached with one or various points to supply internet and connectivity to cellular networks.

The same customary problems of wireless communications, that is, bandwidth limitations, battery power, enhancement of transmission quality and coverage problems have been demonstrated in these networks.

A Wi-Fi network in common includes a set of mobile serves which connect to other mobile serves either straight or via an entry way (base station) (Begam & Murugaboopathi, 2013).

As described by Chavda and Nimavat (2013), wireless networks use some sort of stereo wavelengths in air to deliver and get information instead of using some physical wires. Wireless networks are established by wireless routers and serves.

As a study of Ullah and Rehman (2010), Wi-Fi network is a network in which, PC devices exchanges information with each other without any cable. The interaction method between the PC devices is Wi-Fi. When a PC network wants to connect with another network, the location network must sets within the radio range of each other. Customers in Wi-Fi networks transfer and take the information using electromagnetic waves. Lately, Wi-Fi networks are getting more and more well-known because of its flexibility, convenience, and very cost-effective and cost saving set up.

Wireless networks are getting well known due to their convenience of use. Consumer/user is no more reliant on wires where he/she is simple to shift and enjoy being linked with the network (Ullah & Rehman, 2010).

One of the features of Wi-Fi network that makes it amazing and recognizable among the conventional wired networks is flexibility (Vincent & Meshach, 2012). This function gives customer the ability to shift easily while being linked with the network. Wireless networks relatively simple to set up then wired network. There is nothing to fear about taking the cables/wires in wall and roofs. Wireless networks can be designed according to the need of the customers. These can have client from few of users to large full facilities networks where the client of users is in countless numbers (Vincent & Meshach, 2012). In addition, wireless networks are very useful for places where the cable cannot be set up like hilly places.

On the base of coverage area the wireless network can be separated into:

1. Personal Area network
2. Local Area Network
3. Wide Area Network

1. Personal Area Network
   Communication between computer devices nearby one person has been used as personal area networks (Ullah & Rehman, 2010). Some of the personal area networks are ZigBee, Bluetooth, and Sensor networks. Bluetooth is low cost wireless connection that can

connect devices. These devices usually work within 10 m, with access rate up to 721 Kbps. This technology is greatly used in a range of devices like computer and their accessories like mouse, keyboard, personal digital staff (PDAs), printers, mobile phones, etc. It needs to know that Bluetooth as Wireless Personal Area Network is not 802.11 wireless as it does not execute same job, rather used as replacement for cable in order to link devices. Bluetooth works at 2.4 GHz band, and this may cause interference with wireless LAN instrumentations (802.11b, 802.11g).

2. Local Area Network
   WLAN is standardized by Institute of Electrical and Electronics Engineer (IEEE). In local area networks the users transfer information with each other in local coverage region like building or a campus. WLANs are a replacement of the conventional wired LANs. WLAN is wireless intermediate that is shared by the devices within the WLAN.
   WLANs have earned a great amount of popularity. Because of their mobility features, they are implemented in mobile devices like laptop, PDAs, mobile cell phones, etc. In WLAN, Wireless Ethernet Protocol, IEEE 802.11 is used. WLAN is chiefly used for connecting to the internet. Compared with the wired LAN the information speed of WLAN is so low between 11 and 54 Mbps which wired LAN operates at 100–1000 Mbps. So that, it is better to work with wired networks rather than wireless networks for activities that require high bandwidth.
3. Wide Area Network
   Wireless Wide Area Networks (WWAN) cover geographically larger region than local area networks. The wide area networks almost comprise of one or two local area networks. For instance, satellite systems, paging networks, 2G and 3G mobile cellular are some of WWAN (Fig. 2.1).

### 2.5.1 IEEE Standard for Wireless Networks

IEEE determines the standards for related technologies. Three main operational standards for wireless LAN: IEEE 802.11a, 802.11b, 802.11n and 802.11g have been determined by IEEE. The entire three standards are part of IEEE 802.11 protocol family. The 802.11a standard was ratified in 1999 by IEEE. The 802.11 has a nominal data rate of 54 Mbps, but the actual data rates modifies between 17 and 28 Mbps.

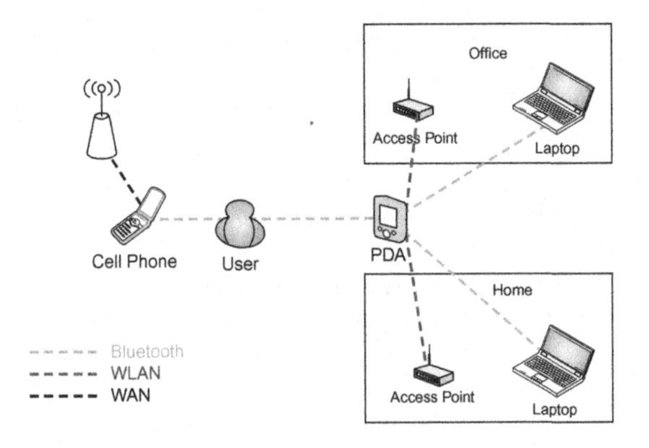


*Figure 2.1 Communications in wireless networks (Ullah & Rehman, 2010).*

802.11b is the most established and frequently deployed wireless network standard. This standard has been used by most of the public wireless "hotspots". It acts in 2.4 GHz spectrum, and the nominal data transfer is 11 Mbps. Almost 4–7 Mbps is the actual data transmission rate achieved by this standard.

## 2.5.2 Categorization of Wireless Networks

Network Elements in a Wi-Fi network connect with each other using Wi-Fi networks. The use of Wi-Fi networks has become more and more well known. In accordance with the kind of network facilities used for interaction, Wi-Fi interaction network are classified into two types: infrastructure networks and infrastructure-less networks (Bakshi, Sharma, & Mishra, 2013).

### 2.5.2.1 Infrastructure Networks

A facilities network consists of Wi-Fi mobile nodes and one or more joins, which connect the Wi-Fi network to the wired network as shown in Fig. 2.2. These joins are known as platform stations. A mobile node within the network searches for the closest platform station, joins to it, and conveys with it.

### 2.5.2.2 Infrastructure-Less Networks

In comparison to facilities networks, each node in this network functions both as a wireless router and a client. The network topology is powerful because the connection among the nodes may differ eventually due to node upgrades. There is no platform place or entry way. Nodes can connect with each other by developing a multiple wish

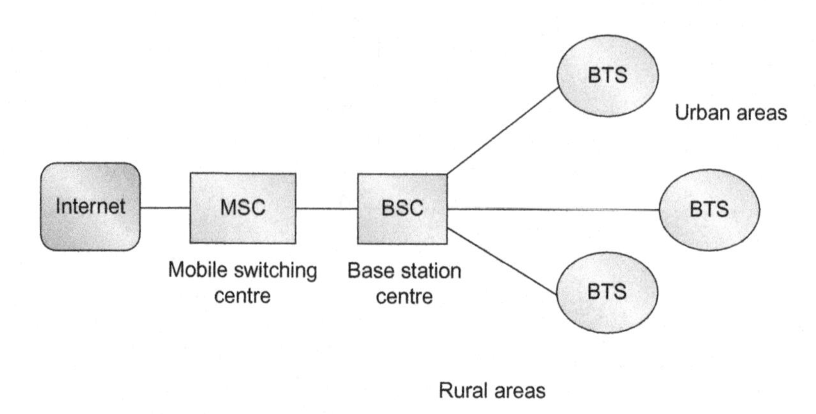


*Figure 2.2 Infrastructure-based network (Bakshi et al., 2013).*

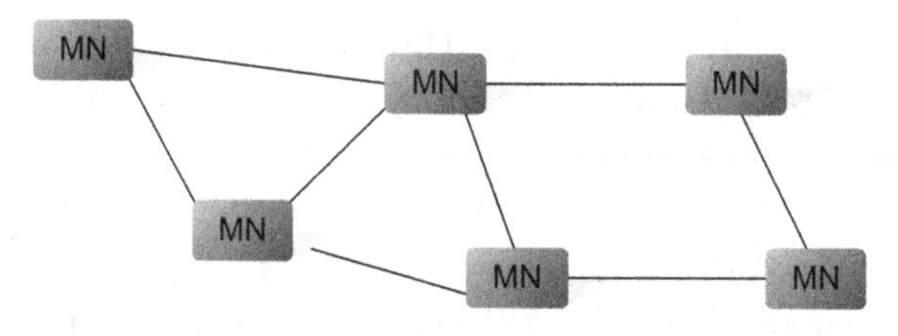


*Figure 2.3 Infrastructure-less network (Bakshi et al., 2013).*

path as proven in Fig. 2.3. Hence there is a need for effective routing method to allow the nodes to connect over multihop tracks without entry way. Since these networks cause many complicated issues, there are many issues for analysis and efforts.

### 2.5.3 Benefits of Wireless Networks

Some of the advantages of wireless networks are:

1. Mobile customers are offered with access real-time information even when they are away from their home.
2. Setting up a Wi-Fi network is easy and fast, and it removes the need for taking out the wires through surfaces and roofs.
3. Network can be prolonged to locations which cannot be wired.
4. Wireless networks offer more versatility and adjust easily to changes in the setting of the network.

### 2.5.4 Weaknesses of Wireless Networks

Here has been listed some of the disadvantages of wireless networks:

1. Interference due to weather, other radio wave gadgets, or obstacles like surfaces.
2. The total throughput is impacted when several relationships prevails.

## 2.6 AD HOC NETWORKS

An ad hoc network is an independent network that does not need a preestablished facility (Chun, Shioura, Tien, & Tokuyama, 2013). Chun further proposed that nodes in an ad hoc network are linked by Wi-Fi hyperlinks, and the e-mails between nodes are often obtained by multihop hyperlinks. With improved passions in mobile emails and the guarantee of practical infrastructure-free e-mails the growth of large-scale ad hoc network has attracted a lot of interest and has been a topic of comprehensive analysis. Fig. 2.4 shows a simple ad hoc network.

As described by Student and Dhir (2013) a Wi-Fi ad hoc network is a decentralized type of Wi-Fi network. The network is ad hoc because it does not depend on a preexisting facility, such as wireless routers in wired networks or accessibility points in handled (infrastructure) Wi-Fi networks. Instead, each node takes part in routing by sending details for other nodes, and so the dedication of which nodes ahead details is made dynamically based on the network relationship. An ad hoc network typically represents any set of networks where all gadgets have equivalent position on a network and are totally able to affiliate with any other ad hoc network gadgets in link client. In general, ad hoc networks have become popular because they can provide mobile

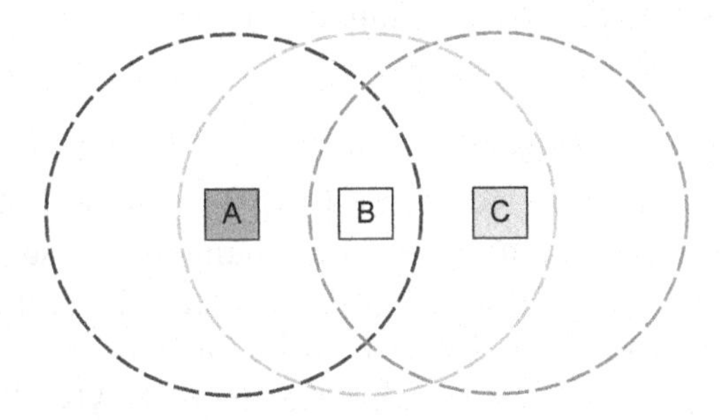


*Figure 2.4 Simple ad hoc network (Singh, Dua, & Das, 2012).*

customers with popular interaction ability, that is, details on accessibility regardless of location (Koyama & Suzuki, 2013).

Gupta and Shrivastava (2013) described that ad hoc networks are built and handled by the component Wi-Fi mobile nodes. The nodes in these networks have restricted transmitting client and link with the rest of the nodes in the network through advanced nodes. Thus account of such an interaction network needs that the component nodes act in the sympathetic manner throughout the function of the network. The nodes in these networks are mobile, resulting in the powerful topology of the network. Due to the participation of advanced nodes in the routing process and the powerful topology of the network, it is necessary that there should be a procedure to find the path between a given couple of source and location time and again. Furthermore, ad hoc networks are Wi-Fi networks where nodes link with each other using multihop hyperlinks Chavda and Nimavat (2013).

As described by Ruzgar and Dagdeviren (2013), ad hoc networks include many independent Wi-Fi gadgets that conveys with each other via stereo alerts. Ad hoc networks have no facilities where the nodes are totally able to be a part of the network and leave the network. The nodes are linked with each other through a Wi-Fi link. A node can work as a router to send the information to the neighbors' nodes. Hence this kind of network is also known as infrastructure-less networks. These networks have no central organization. Ad hoc networks have the abilities to handle any damage in the nodes or any changes that it experiences due to topology changes. Whenever a node in the network is down or leaves the network that causes the link between other nodes interrupts. The affected nodes in the network simply demand for new tracks and new hyperlinks are recognized. Ad-hoc network can be classified in to fixed ad hoc network (SANET) and MANET (MANET) (Ullah & Rehman, 2010).

### 2.6.1 Static Ad Hoc Networks

As described in Ullah and Rehman (2010) research in SANET the geographical location of the nodes or the channels are set. There is no flexibility in the nodes of the networks hence are known as set ad hoc networks. Furthermore, in SANET, roles of gadgets do not change after it signed up with the network, whereas in MANET, gadgets can move randomly (Ruzgar & Dagdeviren, 2013).

## 2.6.2 Mobile Ad Hoc Networks (MANETs)

MANET is an independent system where nodes/stations are connected with each other through wireless links. There is no limitation on the nodes to connect or leave the network, and so the nodes connect or leave freely. Due to the nodes that move freely and the ability of organizing themselves randomly, MANET topology is dynamic that can change fast. The MANETs is unforeseeable from the point of view of scalability and topology because of this characteristic of the nodes (Ullah & Rehman, 2010) (Fig. 2.5).

A MANET is a self-configuring facilities less network of mobile phone gadgets linked by Wi-Fi hyperlinks (Bakshi et al., 2013; Chavda & Nimavat, 2013; Cheng, 2012; Kaur & Rai, 2012; Menon, Johny, Tony, & Alias, 2013; Srivastava, 2012; Student & Dhir, 2013), which reduces their implementation time as well as cost (Bakshi et al., 2013). Ad hoc is Latina and indicates "for this purpose." Each network in a MANET is free to move individually in any route and will therefore change its hyperlinks to other gadgets regularly (Student & Dhir, 2013). Furthermore, each must head traffic irrelevant to its own use and therefore be a wireless router. The main task in building a MANET is outfitting each network to consistently sustain the information required to effectively path traffic (Student & Dhir, 2013). Such networks may function by themselves or may be linked with the bigger online. The growth of notebooks and 802.11/Wi-Fi social media has made MANETs a well-known research subject since the mid-1990s.

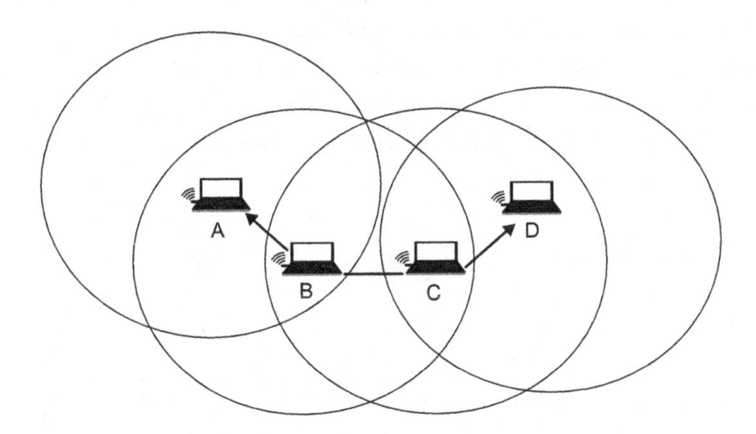


*Figure 2.5 MANET.*

With the developments in Wi-Fi technology and growth of mobile phone gadgets, ad hoc networks will play an important part in allowing existing and upcoming interaction. For both video and information interaction, mobile stereo technology has a knowledgeable fast growth. A MANET is a powerful Wi-Fi network established by a set of mobile serves which connect among themselves through the air without any preexisting facilities. Each node in the MANET can act as a wireless router as well as client. To keep connection in a MANET, all taking part nodes have to execute routing of network traffic (Bakshi et al., 2013).

The success of interaction extremely relies on other nodes collaboration. Therefore, MANET has the property of fast infrastructure-less implementation and no central operator which makes it practical to people and automobiles can thus be internet-worked in places without the preexisting interaction facilities or when the use of such facilities needs Wi-Fi expansion. By increasing client of mobile nodes, ad hoc network facilitates multihop routing by which they can increase the client of Wi-Fi networks (Bakshi et al., 2013). Range relies on the focus of Wi-Fi customers. Fig. 2.6 has illustrated a MANET:

MANET is termed as a Wi-Fi ad hoc network in which nodes are free to shift randomly and mobile nodes can transfer and get the traffic (Dadhania & Patel, 2013). Also mobile nodes can act like wireless routers by sending the others who live nearby traffic to the location node as the wireless routers are multiple hop gadgets (Rafsanjani, Movaghar, & Koroupi, 2008). MANET does not need platform channels of wired facilities. The mobile nodes in Wi-Fi network client can connect with each other because it is a self-organized network. The mobile nodes form a network instantly without a set facilities and main control. The mobile nodes have transmitters and devices with

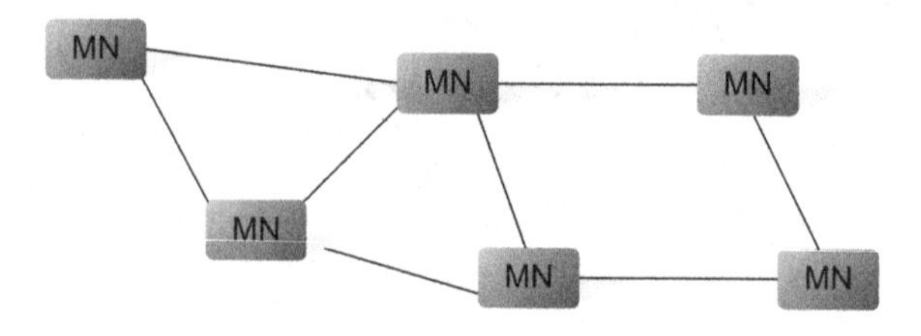


*Figure 2.6 MANET (Bakshi et al., 2013).*

intelligent antennas, which allow the mobile nodes to connect with each other person's (Rafsanjani et al., 2008).

The topology of the network changes whenever by getting in and out of the mobile nodes in the network. In the beginning, MANET was developed for army use but now the MANET is used in many places, such as in catastrophe hit places, information selection in some area, in save tasks, exclusive sessions, and conventions. This idea with ad hoc network makes the full name of MANETMANET. By increasing the network, along with the node flexibility the difficulties of self-configuration of the network become more obvious (Rafsanjani et al., 2008).

Dangore and Sambare (2013) described that a MANET includes mobile serves prepared with Wi-Fi interaction gadgets. The transmitting of a mobile client is obtained by all nodes within its transmitting client due to the transmitted characteristics of Wi-Fi interaction and omni-directional antenna. If two Wi-Fi serves are out of their transmitting varies in the ad hoc networks, other mobile serves situated between them can ahead their information, which successfully develop linked networks among the mobile serves in the implemented place. Due to the flexibility of Wi-Fi serves, each client needs to be prepared with the ability of an independent network or a routing operate without any statically recognized facilities or main control.

In general, nodes help each other in offering details about the topology of the network and discuss the liability of handling the network (Tamilselvan & Sankaranarayanan, 2006). Each mobile node functions as a client when requesting/providing details from/to other nodes in the network, and functions as wireless router when finding and keeping tracks for other nodes in the network.

Also Arunmozhi and Venkataramani (2012) described, a MANET is a selection of mobile phone gadgets that can connect with each other without the use of a predetermined facilities or main control.

MANETs are particularly useful for interaction during mishaps, on the battleground, and conventions. MANETMANETs have obtained improving attention recently due to their flexibility function, powerful topology, and convenience of implementation. A MANET is a self-organized Wi-Fi network which includes mobile phone gadgets, such as notebooks, mobile phones, and PDAs, which can easily shift in the network. In addition to flexibility, mobile phone gadgets act and forward packages for each other to improve the restricted Wi-Fi transmitting client of each node by multihop sending, which is used for various networks, that is, catastrophe comfort, army operate, and urgent emails (Liu, Nishiyama, Ansari, Yang, & Kato, 2013). Furthermore, in MANETs, changes in network topology may dynamically happen in an unforeseen way since nodes have freedom to shift anywhere randomly (Srivastava, 2012).

The network includes a selection of mobile nodes, which simultaneously act as both wireless router and a client, capable of developing immediate network with a powerful topology. Nodes can get into and keep the network easily at any immediate of your energy and effort. Two nodes are said to be linked if they can straight away connect with each other using their receivers. However, interaction is possible with nodes which are outside the immediate achieve of platform place. Mobile nodes which are within the interaction client act as advanced nodes and successfully pass the details via sequence of local trips guaranteeing the connection between the mobile nodes which are outside the immediate protection place of the platform place (Murthy & Manoj, 2004).

#### 2.6.2.1 Categorization of MANETs

When a node wants to transfer information to another node, the destination node should relay within the radio range of the source node that wants to start the communication. The mediatory nodes within the network help in sending the packets from the source node to the destination node. These networks are completely self-organized, having the ability to work anywhere without any infrastructure. Nodes are autonomous and can be router and host at the same time. MANET is self-administering, there is no centralized control, and the communication performs with blind interactive trust between the nodes on each other. The network can be established anywhere without any geographical restrictions. The limited energy resource of the nodes is one of the limitations of MANETs. MANETs are divided to three types:

1. Vehicular ad hoc networks (VANETs)
2. Intelligent vehicular ad hoc networks (InVANETs)
3. Internet-based MANETs (iMANET's)

2.6.2.1.1 Vehicular Ad hoc Networks (VANETs)

A type of MANETs where vehicles are equipped with wireless and form a network without help of any foundation is known as VANET. The equipment is located inside vehicles as well as on the road for gaining access to other vehicles in order to design a network and communication.

2.6.2.1.2 Intelligent Vehicular Ad hoc Networks (InVANETs)

Vehicles form MANET for communication using WiMax IEEE 802.16 and WiFi 802.11. The main goal of planning InVANETs is to avoid vehicle crash so that can keep passengers as safe as possible. This also help drivers to keep secure distance with other vehicles as well as aid them at how much speed other vehicles are approaching. InVANET's applications are also employed for military plans to communicate with each other.

2.6.2.1.3 Internet-Based MANETs (IMANETs)

IMANETs have been used for linking up the mobile nodes and fixed internet gateways. The normal routing algorithms have not been applied in these networks (Ullah & Rehman, 2010).

**2.6.2.2 Features of MANET**

Major features including operating without a central coordinator, Multihop radio relaying, frequent link breakage due to mobile nodes, constraint resources like bandwidth, computing power, battery lifetime, etc. (Dangore & Sambare, 2013). Some main characteristics of MANET are discussed below (Bakshi et al., 2013):

1. Infrastructure less
2. Wireless links
3. Node movement
4. Power limitation
5. Dynamic topologies
6. Self-configuring
7. Bandwidth-constrained and variable capacity links
8. Energy-constrained operation
9. Limited physical security
10. Dynamic topology (Kaur & Rai, 2012)
11. Limited bandwidth (Kaur & Rai, 2012)
12. Energy constrained operation (Kaur & Rai, 2012)
13. Security (Kaur & Rai, 2012)

The MANET has the following typical features (Student & Dhir, 2013):

1. Unreliability of wireless links between nodes.
2. Constantly changing topology.

#### 2.6.2.3 Utilization of MANET

The characteristics of MANET make it more interesting that would bring so many benefits. There are so many research areas in MANET which are under investigation now. One of the important areas is vehicle-to-vehicle communication, where the vehicles would transfer information with each other, keeping a safe distance between them as well as crash warnings to the drivers. MANETs also can be used for automated battlefield and war games. Furthermore, the most important area where MANETs are employed is emergency services such as disaster recovery and relief activities, where traditional wired networks are already damages. Besides, there are so many other application areas such as entertainment, education, and commercial where MANETs are employed for connecting people.

This kind of infrastructure-less network is very useful in situation in which common wired networks is not possible like battlefields, mishaps, etc. Ad hoc social media can be used anywhere where there is little or no interaction facilities or the current facilities is expensive or undesirable to use. Ad hoc social media allows us to product our sustain relationships to the network as well as easily including and eliminating gadgets to and from the network. MANET can be used to quite a number of use cases where traditional social media cannot be used. MANET is used in following places (Bakshi et al., 2013):

1. Military battlefield
2. Sensor networks
3. Disaster area network
4. Personal area network

#### 2.6.2.4 Benefits of MANET

Some of the advantages of MANET are:

1. Router free
2. Mobility
3. Speed
4. Fault tolerance

5. Connectivity
6. Fast installation
7. Cost

**2.6.2.5 Weaknesses of MANET**

MANETs are very flexible for the nodes; this means nodes can freely join and leave the network. There is no main body that can do controlling on the nodes entering and leaving the network. Because of these characteristics of MANET, it is vulnerable to attacks that we discuss it below:

1. Nonsecure boundaries:
   Due to no clear secure boundary, MANET is vulnerable to various kinds of attacks. It's the nature of MANET that nodes have the freedom to join and leave the network. Node can join to the network automatically if the network is in the radio range of the node, and thus it can transfer information with other nodes in the network. As a result of no secure boundaries, MANET is more affected by attacks.
   The attacks on MANET may be passive or active, leakage of information, false message reply, denial of service (DoS), or changing the data integrity. The links are compromised and open to different link attacks. Attacks on the link hinder between the nodes and then attack the link, afterwards demolish the link after executing malicious behavior.
   There is no protection like firewalls against attacks, which result the vulnerability of MANET to attacks. When security compromises, attacks can result to spoofing of node's identity, data tempering, confidential information leakage, and impersonating node (Parsons & Ebinger, 2009).
2. Compromised node:
   Some of the attacks are to get access inside the network in order to get control over the nodes in the network to exploiting them; it means to accomplish their malicious activities. Since mobile nodes in MANET are free to move, join, or leave the network, this means that the mobile nodes are autonomous (Roy, Chaki, & Chaki, 2010), it is very difficult for the nodes to prevent malicious behavior of nodes which they are communicating with. It is very easy for a compromised node to change its position because of ad hoc mobility, therefore often making it more hard and troublesome to

trace the malicious activity. It has been realized that these threats from compromised nodes inside the network is more perilous than attack threats from outside the network.

3. No central management:
   MANET is a self-arranged network, which is including of mobile nodes that communicate with each other without a central control. Each of these nodes acts as a router and thus can forward and receive packets (Shanthi et al., 2009). MANET works without any preexisting foundation. This absence of centralized management causes MANET more insecure to attacks. In ad hoc networks, detecting attacks and monitoring the traffic in highly dynamic and for large scale is very troublesome according to no central management. When there is a central entity, it takes care of the network by applying suitable security; it also authenticates which node can join and which cannot. When nodes connect to each other on the basis of blind mutual trust on each other, a central entity can administer this by using a filter on the nodes to discover the malicious one and let the other nodes know which node is malicious.
4. Problem of scalability:
   In traditional networks, the network is designed and every device is connected to other devices with help of wire. The network topology and the scale of the network, while designing, has been defined and it would not change much during its life. In other words, we can say that the scalability of the network is clarified in the starting phase of the network designing. The case is completely different from the MANET because MANET nodes are mobile and due to their mobility, the scale of the MANETs is changing. It is so hard to know and foretell the numbers of nodes in the MANETs in the future. The nodes are free to move inside and outside the MANET which makes it very scalable and shrinkable. According to this characteristic of MANETs, the protocols and all the services that a MANET provides must be adaptable to such changes.

Limitations of MANET can be listed as below:

1. Bandwidth constraints
2. Processing capability
3. Energy constraints
4. High latency
5. Transmission errors
6. Security
7. Location

8. Roaming
9. Commercially unavailable
10. Limited resources
11. Scalability problems
12. No central check on the network
13. Dynamic topology, where it is hard to find out malignant nodes

## 2.7 ROUTING

Routing is regarded to be one of the significant difficulties experienced by Wi-Fi ad hoc networks (Menon et al., 2013), as the topology is not set and is never stand still and also there is no central management for the network.

Routing is an integral aspect of MANETs as it gives the better choice of tracks. Routing is the procedure of creating the choice of tracks in a network along which network traffic can be delivered from a resource to a location node (Bhushan, Gupta, & Nagpal, 2013; Gupta & Shrivastava, 2013).

## 2.8 AD HOC NETWORK ROUTING PROTOCOLS

An ad hoc routing methods the official set of guidelines or guidelines which manages how nodes choose which way to path packages between computers in a MANET (Bhushan et al., 2013; Menon et al., 2013; Student & Dhir, 2013). In ad hoc networks, nodes are not acquainted with the topology of their networks. Instead, they have to find it. Routing methods gives the requirements for how routers connect with each other; they commonly spread the details that allow them to choose tracks with the help of routing methods between any two nodes on a computer network. Each wireless router contains a priori information only of networks which are straight linked with it. First of all, this detail is distributed by the routing method to immediate others who live nearby and then throughout the network. This way, routers become conscious of the topology of the network (Srivastava, 2012).

The main objective of such a routing method is to make sure appropriate and effective path organization between couple of nodes, so that information is provided in regular basis. Different methods are then analyzed based on evaluation such as the bundle fall amount, the expense presented by the routing method, end-to-end bundle setbacks, network throughput, etc. (Bhushan et al., 2013).

## 2.9 MANETs ROUTING PROTOCOLS

MANET is growing rapidly since past 20 years. The profit in their popularity is because of the easy deployment, infrastructure less, and their dynamic nature. MANETs made a new set of demands to be executed and to provide a better efficient end-to-end transmission. MANETs work on TCP/IP structure to provide the means of transmission among transmitting work stations. Work stations are mobile and they have limited resources, and so the traditional TCP/IP model needs to be refurbished or changed in order to compensate the MANETs mobility to provide efficient functionality. Hence the key research area for the researchers is routing in any network, Routing protocols in MANET are challenging and interesting manners, researchers are giving huge amount of attention to this key area.

### 2.9.1 Categorization of Routing Protocols

A huge number of methods have been designed for ad hoc mobile networks; working with the restrictions of mobile networks like great energy intake, low data transfer usage, and great mistake prices.

Most of the methods provided for MANET are either practical or sensitive methods. The methods are further classified into two categories (Bhardwaj, Sharma, & Dubey, 2012; Natarajan & Mahadevan, 2013; Singh & Sharma, 2012): (i) proactive (or on-demand) routing methods and (ii) reactive (or table-driven) routing protocols. Fig. 2.7 shows this classification of MANET routing protocol:

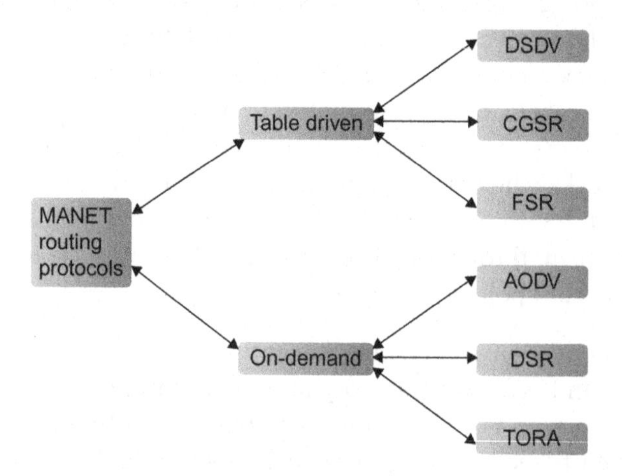


*Figure 2.7 MANET routing protocols (Natarajan & Mahadevan, 2013).*

Multiple routing methods, such as ZHLS and ZRP, etc., is a new generation of methods, which brings together the features of both on-demand and table-motivated routing methods.

After generating this new routing method, routing methods generally be categorized as (Bhushan et al., 2013; Kaur & Kumar, 2013; Kaur & Rai, 2012; Singh et al., 2012; Srivastava, 2012): (i) practical (or on-demand) routing methods, (ii) sensitive (or table-driven) routing methods, and (iii) multiple routing methods. Student and Dhir (2013) categorized routing methods into eight parts while is described as follows:

**2.9.1.1 Table-Driven (Proactive) Routing:**

Practical routing methods are also known as table-motivated routing methods. These methods have their capability to sustain routing platforms that shop details regarding the tracks from one node in the network to relax of other nodes (Bhardwaj et al., 2012; Bhushan et al., 2013; Dhaigude et al., 2012; Edward & Seethalakshmi, 2012; Kaur and Kumar, 2013; Kaur and Rai, 2012; Srivastava, 2012; Natarajan & Mahadevan 2013). Here, all nodes upgrade their platforms to protect interface by trading routing details between the taking part nodes. Every node in the network knows about the other node in advance, in other terms, the whole network is known to all the nodes creating that network (Kaur & Kumar, 2013). All the routing details are usually kept in platforms (Kaur & Kumar, 2013). Some of the current proactive routing methods are DSDV, optimized link state routing protocol (OLSR), and Wi-Fi routing method (Kaur & Kumar, 2013).

There is little time-wait (time invest in direction development process) occurs (Dhaigude et al., 2012), So a quickest direction can be discovered without spending more time; however, these methods are not appropriate for very heavy ad hoc networks due to great traffic may occur. Several variations of proactive methods have been suggested for eliminating its drawbacks and use in ad hoc networks. It preserves the unicast tracks between all couple of nodes without considering whether all tracks are actually used or not. It can be of two kinds based on the methods which have been proven in the next area. Furthermore, it uses link condition routing criteria. The proactive methods are not appropriate for a huge network because each and every node preserves all details of every node in routing table (Bhardwaj et al., 2012). So this protocol is use as powerful network (eg, OLSR).

The benefit of these methods is that a resource node does not need route-discovery techniques to discover a direction to a location node. However, the disadvantage of these methods is that keeping a regular and up-to-date routing table needs significant texting cost, which requires information transfer usage and energy and reduces throughput, especially in the situation of a huge client of great node flexibility (Jacob & Seethalakshmi, 2012). The primary drawbacks of such methods are:

- Specific quantity of information for servicing.
- Slow response on reorienting and problems.

Some of the applications of proactive routing protocols included the table-driven technique is being used for different upgrading hyperlinks and also it can use both the range vectors and link statuses as used in the set networks (Natarajan & Mahadevan, 2013). This function although useful for datagram traffic, happens upon significant signaling traffic and energy intake (Kaur & Rai, 2012). Practical methods are not appropriate for huge networks as they need to sustain node records for each and every node in the routing table of every node (Kaur & Rai, 2012).

Proactive routing protocols work in a different way compared with reactive routing protocols. These protocols continuously preserve the updated topology of the network. Each node in the network knows about other nodes in advance; in other words, the whole network is known to all the nodes created that network. Usually, some tables held all the routing information (Abolhasan, Wysocki, & Dutkiewicz, 2004). Any time there is a change in the network topology, according to the change, these tables are updated. The nodes interchange topology information with each other; at any time they need route information, they can have it (Abolhasan et al., 2004).

#### 2.9.1.2 Reactive (On-Demand) Routing

Reactive protocols are also known as on-demand driven reactive protocols. These protocols do not start route discovery by themselves, until it has been made a request to, when a source node asks for finding a route (Bhardwaj et al., 2012; Bhushan et al., 2013; Kaur & Kumar, 2013; Kaur & Rai, 2012; Natarajan & Mahadevan, 2013; Srivastava, 2012). That is why these techniques are known as reactive techniques. These protocols install routes when required. When a node wants to connect with another node in the network and the source node does not have a path to the node it wants to connect with, reactive routing protocols will set up a path for the source to the destination node. Normally reactive protocols

1. Do not discover path until required
2. Uses flooding strategy to spread the inquiry, when attempts to discover the destination on-demand.
3. Do not use up bandwidth for delivering information.
4. They use up bandwidth only, when the node start transferring the information to the destination node.

Some of the most used on requirement routing techniques are DSR, AODV, and entrance control allowed on requirement routing method. The primary benefits of using this method is that information transfer usage is being used successfully (Natarajan & Mahadevan, 2013). In contrast, the primary drawbacks of such methods are:

1. High-latency time in path finding.
2. Extreme surging can lead to network blocking.
3. Flow focused routing
4. Hybrid (both proactive and reactive) routing

As Kaur and Kumar (2013), Kaur and Rai (2012), and Singh et al., (2012) describes, both of the practical and reactive routing techniques have some benefits and drawbacks. In multiple routing a mixture of practical and reactive routing techniques are used which are better than the both used in solitude. It contains the key benefits of both techniques. As an example, accomplish the reactive routing protocol such as AODV with some practical features by relaxing tracks of effective locations which would definitely decrease the wait and expense, so renew period can enhance the performance of the network and node. These techniques can integrate the service of other techniques without limiting with its own benefits. Illustrations of multiple techniques are area routing method, obscure spotted link state.

#### 2.9.1.3 Hybrid Routing

Hybrid protocols utilize the capabilities of both reactive and proactive protocols, and unite them together to achieve better results. The network is separated into zones and use different protocols in two different zones that one protocol is used within the zone and other protocol is used between them. An example of hybrid routing protocol is zone routing protocol (ZRP). For route constitution within the nodes neighborhood ZRP uses proactive mechanism, it takes the advantages of reactive protocols for communication between the neighborhoods. These local neighborhoods are specified as zones, and thus the protocol is called as ZRP. Every zone may have various sizes, and each node can be within multiple overlapping zones. The size of zone has been given by radius of length P, where P is number of hops to the perimeter of the zone (Ullah & Rehman, 2010).

Hybrid protocol is appropriate for huge networks where huge client of nodes are present (Kaur & Rai, 2012). This huge network is separated into a set of areas where routing inside the zone is conducted by using reactive strategy and outside the zone is done using reactive strategy. Fig. 2.8 shows another classification of MANET routing protocol that included hybrid protocol. There are various popular multiple routing techniques for MANET like ZRP, SHRP. These are practical, reactive, multiple, and place-based routing (Dhaigude et al., 2012).

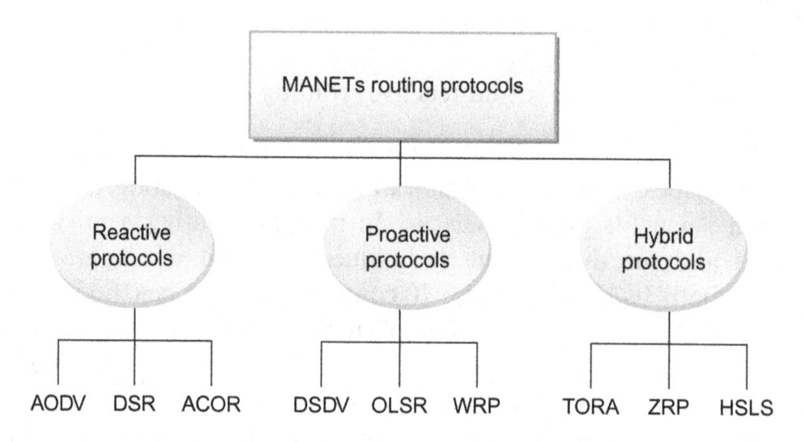


*Figure 2.8 Another classification of MANET routing protocols (Kaur & Kumar, 2013).*

## 2.10 OPTIMIZED LINK STATE ROUTING PROTOCOL (OLSR)

The OLSR protocol is explained in RFC3626 (Ullah & Rehman, 2010). OLSR is a proactive routing protocol that is also known as table-driven protocol that updates its routing tables. Three types of control messages of OLSR are described below.

1. Hello
   This control message is transferred for sensing the neighbor and for multipoint distribution relays (MPRs) calculation.
2. Topology control (TC)
   These are link state signaling that is executed by OLSR. MPRs are used to improve the efficiency of these messaging.
3. Multiple interface declaration (MID)
   MID messages include the list of all IP addresses used by any node in the network. All the nodes running OLSR send these messages on more than one interface.

### 2.10.1 OLSR Working

#### 2.10.1.1 Multipoint Relaying (MPR)

OLSR spreads the network topology information by flooding the packets throughout the network. The flooding works in such a way that every node that received the packets retransmits the received packets. These packets include a sequence number so as to keep away from loops. The receiver nodes record this sequence number to be sure that the packet is retransmitted once. The basic concept of multipoint relaying (MPR) is to decrease the duplication or loops of retransmissions of the packets.

Only MPR nodes transmit route packets. The nodes within the network save a list of MPR nodes. MPR nodes are chosen within the vicinity of the source node. Choosing of MPR is depended on HELLO message sent between the neighbor nodes. The choosing of MPR is such that a path exists to each of its 2 hop neighbors through MPR node. Routes are set up, once it finished the source node that wants to begin transmission can start sending information (Fig. 2.9).

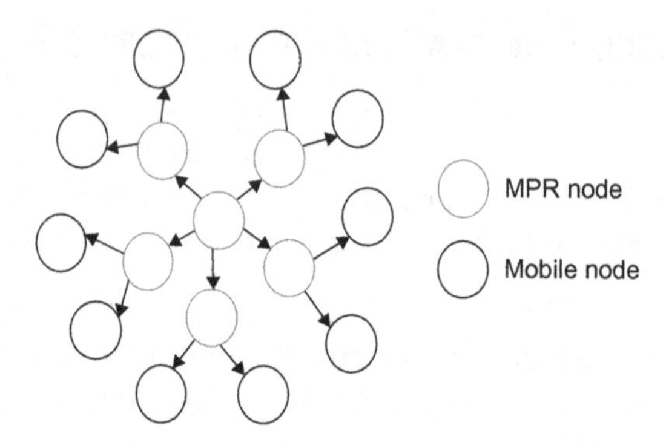


*Figure 2.9 Flooding packets using MPR.*

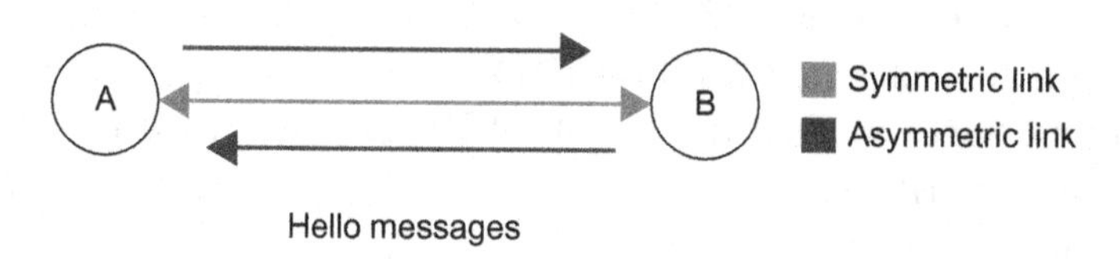


*Figure 2.10 Hello message exchange.*

By looking at Fig. 2.10 below, the whole process can be understood. The nodes displayed in the figure are neighbors. Node "A" sends a HELLO message to the neighbor node "B". When node B receives this message, the link is asymmetric. The same case happens when node B sends HELLO message to node A. When there is two communication ways among both of the nodes, we call the link as symmetric link. HELLO message includes all the information about the neighbors. MPR node broadcast TC message, together with link status information at a predetermined TC interval.

## 2.11 AD HOC ON DEMAND DISTANCE VECTOR ROUTING PROTOCOL (AODV)

AODV is a sensitive method that responds at will. This is the alteration or an enhancement of DSDV (Chaba et al., 2009; Dadhania and Patel, 2013; Edward & Seethalakshmi, 2012; Kumar & Sharma, 2013; Natarajan & Mahadevan, 2013). It allows multihop, self-starting, and powerful routing in network (Kumar & Sharma, 2013). AODV never generates circles as there cannot be any cycle in the routing table of any node because of the idea of series client reverse obtained from DSDV (Kumar & Sharma, 2013).

In AODVs, protocol development procedure is began by the node that wants to connect with the other node and for that recommend, it shows a HELLO idea after a particular time frame, thus a node keeps paths of only its next hop (Kumar & Sharma, 2013). Whenever a node want to connect with a node that is not its next door neighbor, it basically transmitted route request message (RREQ) massage that contain RREQ ID, location IP deal with, location sequence number, resource IP deal with, resource sequence number, and hop depend (Kumar & Sharma, 2013).

It is designed to reduce needing network-wide, while showing it excessive. AODV does not sustain paths from every node to every other node in the network rather they are found as and when needed and are managed only provided that they are needed (Gupta & Shrivastava, 2013).

AODV is explained in RFC 3561 (Ullah & Rehman, 2010). Its reactive protocol, when a node wishes to start communication with another node in the network to which it has no route; AODV will provide topology information for the node. AODV employ control messages to discover a route to the destination node in the network. There are three kinds of control messages in AODV that are discussed as following:

1. RREQ:
   When source node needs to communicate with another node in the network, it sends RREQ message. AODV broadcasts RREQ message, using spreading out ring technique. In every RREQ message, there is a time-to-live (TTL) value; the value of TTL expresses the number of hops the RREQ should be sent.
2. Route reply message (RREP):
   A node that have a requested identity or any mediatory node that has a route to the requested node creates an RREP message and sends it back to the originator node.
3. Route error message (RERR):
   During active routes, each node in the network keeps monitoring the link status to its neighbor's nodes. When the node discovers a link crack in an active route, RERR message is created by the node in order to inform other nodes that the link is down.

### 2.11.1 Routing in AODV

AODV can cope with any type of flexibility rate and a wide range of information traffic. The two main networks used by the AODV method to set up and sustain the relationship between any couple of nodes are as follows (Hassnawi, Ahmad, Yahya, Aljunid, & Elshaikh 2012)

1. Route discovery mechanism.
2. Route maintenance mechanism.

Fig. 2.11 reveals the mechanism of the AODV routing protocol. In Fig. 2.12, it displays how RREQs flood in AODV. Fig. 2.13 demonstrates route reply in AODV.

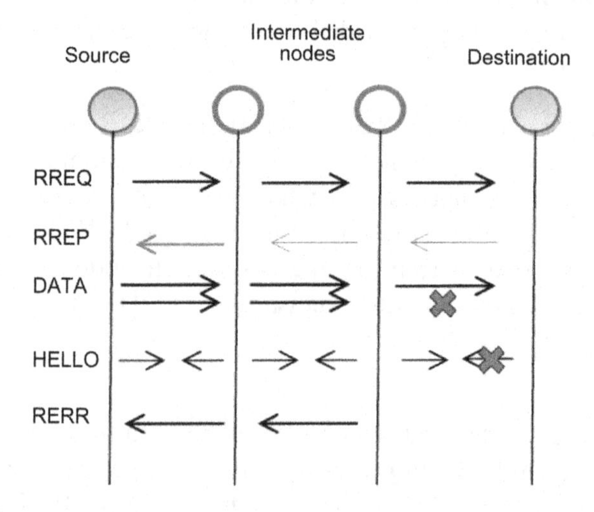


*Figure 2.11 AODV mechanisms (Hassnawi et al., 2012).*

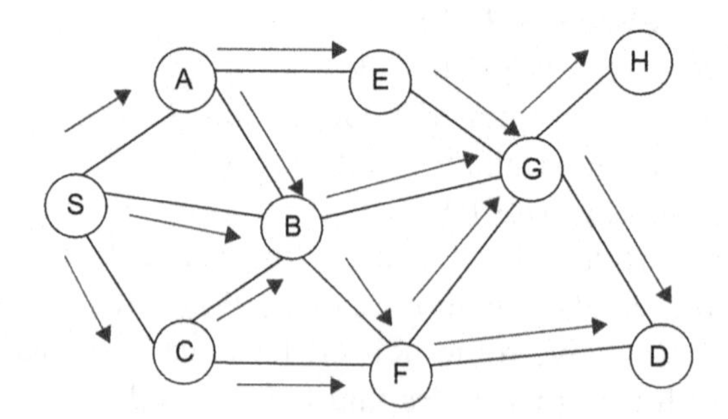


*Figure 2.12 Flooding RREQ in AODV (Bhushan et al., 2013).*

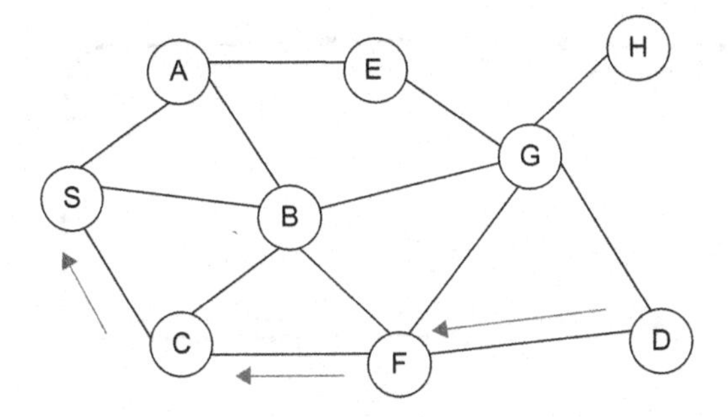


*Figure 2.13 Route reply in AODV (Bhushan et al., 2013).*

### 2.11.1.1 Route Discovery Mechanism in AODV

If a sender (source node) needs a direction to a location, it shows a route request (RREQ) concept. Every node also preserves a transmitted ID which, when taken together with the originator's IP deal with, exclusively recognizes an RREQ. Whenever a sender broadcasts an RREQ, it amounts its transmitted ID and series client by one.

The sender buffers this RREQ for path discovery time so that it does not reprocess it when its others who live nearby deliver it returning. The sender then stays for net traversal time (NETT) for an RREP. If an RREP is not obtained within now, the sender will rebroadcast another RREQ up to RREQ TRIES periods. With each extra effort, the patiently waiting time (NETT) is more than doubled. When a node gets an RREQ concept, it has not seen before, it places up an opposite direction returning to the node where the RREQ came from. This opposite direction has a lifetime value of active route timeout (Marti, Giuli, Lai & Baker, 2000). The opposite direction access is saved along with the details about the asked for location deal with. If the node that gets this concept does not have a direction to the location, it rebroadcasts the RREQ. Each node keeps a record of the client of trips, the concept has created, as well as which node has sent it the transmitted RREQ. If nodes get an RREQ which they have already prepared, they eliminate the RREQ and do not forward it. If a node has a direction to the location, it then responds by unicasting an RREP returning to the node, it obtained the request from. The answer then is sent returning to the sender via the opposite direction set by the RREQ. The RREP originates returning to the source; nodes set up and forward the suggestions to the location. Once the resource node gets the RREP, the direction has been recognized and the

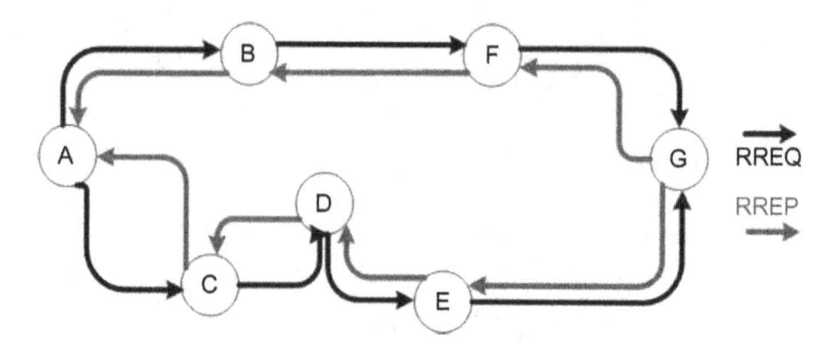


*Figure 2.14 AODV route discovery.*

resource begins to deliver details packages to the location. AODV contains a marketing procedure to manage the RREQ overflow in the direction discovery procedure. It uses a growing band look for originally to discover tracks to an unidentified location. In the growing band look for, progressively bigger communities are explored to discover the location. The look for is managed by the TTL area in the IP headlines of the RREQ packages. If the direction to a formerly known location is required, the before hop-wise range is used to improve the look for.

When a node "A" wants to start communication with another node "G" as displayed in Fig. 2.14, it will create an RREQ. This message is spread by a limited flooding to other nodes. This RREQ control message forwards to the neighbors, and those nodes also forward the control message to their neighbors' nodes. This process of finding destination node continues until it finds a node that has a fresh sufficient route to the destination or destination node is located itself. Once the destination node is located or a mediatory node with sufficient fresh routes is located, they produce control RREP to the source node. When RREP has been received by the source node, a route is set up among the source node "A" and destination node "G". Once the route is set up between "A" and "G", node "A" and "G" can communicate with each other. Fig. 2.14 portrays interchanging of control messages among the source node and destination node.

#### 2.11.1.2 Route Maintenance Mechanism in AODV

As described in Hassnawi et al. (2012), the part of path servicing is to offer reviews to the sender in situation a link damage happens, to allow the path to be customized or rediscovered. A path can leave the workplace basically because one of the mobile nodes has shifted. If a resource node goes, then it must uncover a new path. If an advanced node goes, it must notify all its others who live nearby that may need this hop. This new concept is sent to all the other trips and the old path is removed. The resource node must then uncover a new path or else the node upstream of that crack may select to fix the link regionally if the location was no further than MAX_REPAIR_TTL trips away. To fix the link crack, the node amounts the series client for the location and then shows an RREQ for that location. One suggested way for a node to keep a record of its others who live nearby is by using HELLO details. These are regularly sent to identify link problems. Upon getting alert of a damaged link, the resource node can reboot the rediscovery procedure. If there is a link damages, a route error (RERR) concept can be transmitted on the net. Any client that gets the RERR invalidates the path and rebroadcasts the big mistake details with the inaccessible location details to all nodes in the network.

When there is a link down or a link between nodes is broken that causes one or more than one links inaccessible from the source node or neighbor's nodes, the RERR message sends to the source node. For instance, as it is displayed in Fig. 2.15, when RREQ message is broadcasted for locating the destination node that in the figure is from the node "A" to the neighbors nodes, at node "E" the link is broken between node "E" and node "G", so an RERR message is produced by the node "E" and sent to the source node notifying it a route error is happened, where "A" is source node and "G" is the destination node.

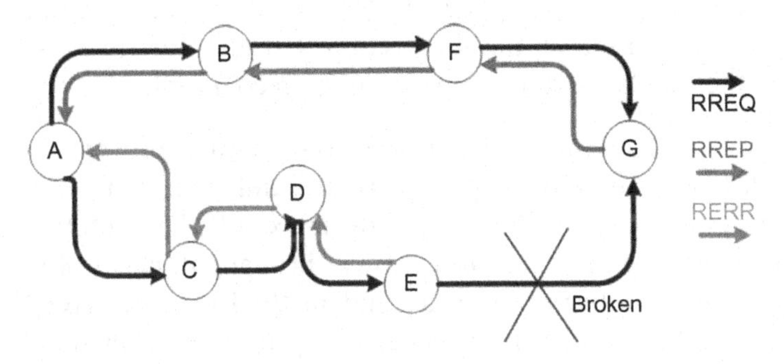


*Figure 2.15 Route error message in AODV.*

### 2.11.2 Features of AODV

AODV routing protocols can be specified by some specifications like:

1. Unicast, broadcast, and multicast communication.
2. On-demand route establishment with small delay.
3. Multicast trees connecting group members maintained for lifetime of multicast group.
4. Link breakages in active routes efficiently repaired.
5. All routes are loop-free through use of sequence numbers.
6. Use of sequence numbers to track accuracy of information.
7. Only keeps track of next hop for a route instead of the entire route.
8. Use of periodic HELLO messages to track neighbors (Singh et al., 2012).

### 2.11.3 Benefits and Weaknesses of AODV

The advantages of AODV method (Gupta & Shrivastava, 2013) are that it prefers the least crowded path instead of the quickest path and it also facilitates both unicast and multicast bundle signals even for nodes in continuous activity. It also reacts very easily to the topological changes that impact the effective tracks. AODV does not put any extra running costs on information packages as it does not make use of resource routing. The restriction of AODV method is that it expects/requires that the nodes in the transmitted method can identify each other's shows. It is also possible that a real path is terminated and the dedication of an affordable expiration time is difficult. The reason behind this is that the nodes are mobile and their delivering prices may vary commonly and can modify dynamically from node to node. Moreover, as the dimension network develops, various performance analytics begin reducing. AODV is susceptible to various types of attacks as it based on the supposition that all nodes must work and without their collaboration no path can be recognized.

It can be determined from the trial results proven here that under different automobile prices of rate and with a different client of information resources, the performance of the AODV routing method is good, with regard to network throughput, client of packages missing, bundle transmitting rate, and method expense. AODV may be regarded "the best," given its capability to sustain relationships by regular return of information. AODV provided almost all packages even at different automobile rate principles.

## 2.12 DYNAMIC SOURCE ROUTING PROTOCOL (DSR)

Dynamic source routing protocol shortened as DSR is also a reactive protocol. By using update its route caches DSR finds new routes. When a new route is detected or when there is a direct route between source and destination node, DSR updates its cache. When a node wants to transfer data, it specifies a route for the transferring and then initiates transferring data through the specified route. There are two processes in DSR for route discovery and maintenance which are explained as following:

### 2.12.1 Route Discovery Process

When a source node wants to start communication with another node in the network, it checks its routing cache. If there is no route available to the destination in its cache or a route is expired, it sends RREQ. When the destination is located or any mediatory node that has fresh sufficient route to the destination node, RREP produces (Zhu, Lee, & Saadawi, 2003). When RREP has been received by the source node, it updates its caches and the traffic is sent through the route.

### 2.12.2 Route Maintenance Process

When transferring of data begins, the node that is transferring data has responsibility to confirm the next hop that will receive the data along with source route. If the node does not receive any confirmation to the originator node, it produces a route error message. And thus the originator node again will do new route discovery process.

## 2.13 SECURITY CHALLENGES IN MANETs

The most important issue for the basic functionality of network in MANET is security. Availability of network services, confidentiality, and integrity of the information can be gained by pledging that security concerns have been solved. Because of MANET's characteristics like open medium, changing its topology dynamically, lack of central monitoring and management, cooperative algorithms and no clear defense mechanism, these networks often suffer from security attacks. The battle field position for the MANET versus the security threats have been modified by these elements.

Recently, security of computer networks has been of serious matter which has extensively been argued and drew up. The arguments mostly contained only static and networking based on wired systems. And thus, MANET security still needs more considerations and progression (Biswas & Ali, 2007). With the appearance of ongoing and new paths for networking, new complications and concerns rise for the essentials of routing. In comparison with the wired network, MANET is different. The routing protocols that are majorly constructed for internet are different from the MANETs. Traditional routing table was mostly made for the hosts which are connected wired to a nondynamic backbone (Pegueno & Rivera, 2006). So they cannot support ad hoc networks principally considering the movement and dynamic topology of these networks.

The routing protocols are vulnerable to different attacks according to several elements including lack of infrastructure, absence of already established trust relationship among the different nodes, and dynamic topology (Lu et al., 2009). Main vulnerabilities that have been so far acquired are mainly types which contain selfishness, dynamic nature, and severe resource restriction and also open network medium. In spite of the above protocols, there are attacks in MANET which can be classified in passive, active, internal, external and network-layer attacks, routing attacks and packet forwarding attacks.

MANET works without a centralized administration where node communicates with each other on the base of mutual trust. This quality makes MANET more vulnerable to be utilized by an assailant from inside the network. Wireless links also make the MANET more susceptible to assaults which make it easier for the assailant to crevasse inside the network and reach to the ongoing communication (Biswas & Ali, 2007; Vinayakray-Jani, 2002). Mobile nodes' existence within the extent of wireless link can overhear and even be involved in the network.

### 2.13.1 Categorizations of MANET Attacks

The attacks can be classified on the foundation of the source of the attacks, that is, internal or external, and on the action of the attack, that is, passive or active attack. Because the aggressor can utilize the network either as internal, external or/as well as active or passive attack against the network, this categorization is important.

To allow the purpose of a protected network, you must understand different kinds of attacks identify them and correctively surpass those (Agrawal et al., 2010). They are categorized in two main kinds. Bhattacharyya, Banerjee, Bose, Saha, and Bhattacharya (2011) categorized the current attacks into two wide categories: DATA traffic attacks and CONTROL traffic attacks. This classification is depending on their common features and attack objectives. For example, black hole attack drops packages whenever, while gray-hole attack also falls packages but its activity is depending on two conditions: time or sender node. But from network perspective, both attacks fall packages and gray-hole attack can be regarded as a black hole attack when it begins losing packages. So they can be categorized under a single classification. There are few attacks that have effects on both DATA and CONTROL traffic, so they cannot be categorized into these groups easily (Bhattacharyya et al., 2011; Bhushan et al., 2013). In Fig. 2.16, we can see this classification of MANET attacks:

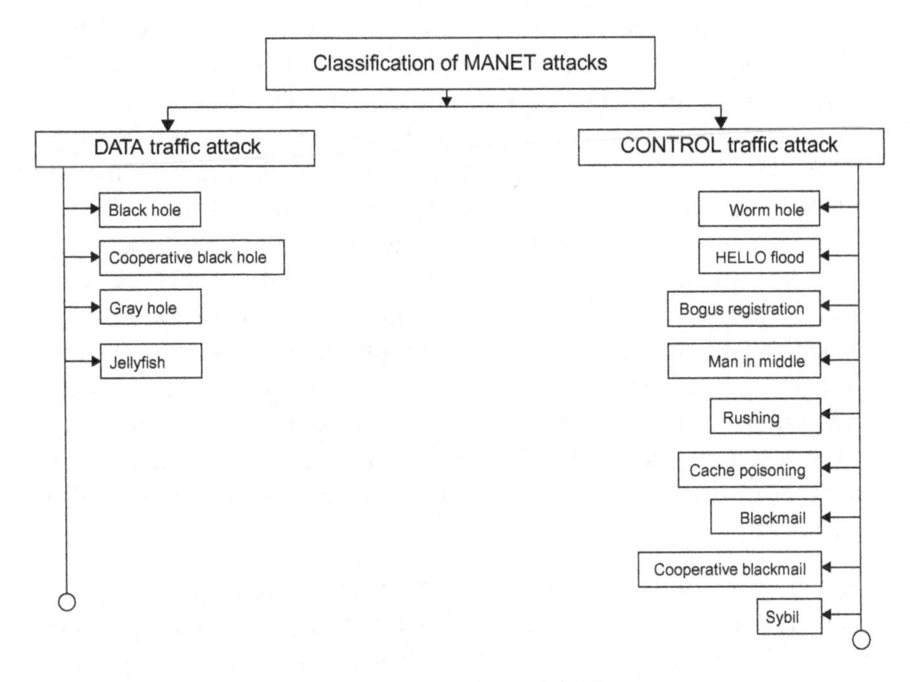


*Figure 2.16 Classification of MANET attacks (Bhattacharyya et al., 2011).*

1. DATA traffic attack:
   DATA traffic attack offers either in nodes losing data packages moving through it or in postponing of sending of your details packages. Some types of attacks choose sufferer packages for losing while some of them fall all of them regardless of sender nodes. This may extremely break down the service quality and increase end-to-end wait. This also causes important loss of important data (Bhattacharyya et al., 2011).
   Some of the other attacks in this classification include gray-hole attack (Bhattacharyya et al., 2011; Vishnu & John, 2010), jellyfish attack.
2. CONTROL traffic attack:
   MANET is naturally susceptible to attack due to its essential features, such as start method, allocated nodes, independence of nodes contribution in network (nodes can be an aspect of and keep the network on its will), deficiency of central power which can implement protection on the network, allocated coordination and collaboration. The current routing methods cannot be used in MANET due to these factors (Bhattacharyya et al., 2011).
   Many of the routing methods developed for use in MANET have their personal attribute and guidelines. Two of the most commonly used routing methods is AODV, which depends on personal node's collaboration in developing a real routing table and powerful MANET on-demand, which is a quick light-weight routing method developed for multiple hop networks. But each of them is depending on believed in on nodes playing network. The first thing in any effective attack needs the node to be part of that network. As there is no restriction in becoming a member of the network, harmful node can be an aspect of and interrupts the network by hijacking the routing platforms or skipping legitimate tracks. It can also eavesdrop on the network if the node can set up itself as the quickest path to any location by taking advantage of the unsafe routing methods. Therefore it is essential that the routing method should be as much protected as it can be.

Some of the other attacks in this classification are: earthworms hole attack (Bhattacharyya et al., 2011; Jhaveri et al., 2010), man in center attack (Bhattacharyya et al., 2011), blackmailing and co-operative blackmailing attack (Bhattacharyya et al., 2011), etc. Though there can be other types of attack, such as performing attacks, which is not management attack, they can be handled as an aspect of actual part protection methods. In another research performed by Saha et al. (2012), MANET attacks have been classified as following:

1. Mobile vs. Wired attackers (Saha et al., 2012):
   Mobile assailants have the same abilities as that of the other nodes of any particular ad hoc network. Having the same resource restrictions, their abilities to harm the networks functions get also restricted. For example, with the restricted transferring abilities and battery abilities, mobile assailants could only jam the wireless links within its area. They are not able to release the network performing attacks to affect the whole networks functions. On the other hand, wired assailants are assailants that are able of accessing the exterior sources such as the power. Since they have more sources, they could release more severe attacks in the networks, such as performing the whole networks or breaking expensive cryptography methods. Lifestyle of the wired assailants in the ad hoc networks (especially in the open environment networks) is always possible as long as the wired assailants are able to locate themselves in the interaction range and have accessibility of the wired infrastructures.
2. Passive versus active attacks (Saha et al., 2012):
   Classes of attack might include passive tracking of e-mails, active network attacks, close-in attacks, exploitation by associates, and attacks through the company. Passive attack is an inactive attack watches unencrypted traffic and looks for clear-text security passwords and delicate details that can be used in other types of attacks.
   Passive attacks include traffic research, tracking of unsecured e-mails, decrypting weakly secured traffic, and catching verification details such as security passwords. Passive interception of network functions enables enemies to see future activities. Passive attacks result in the disclosure of details or information to an enemy without the approval or knowledge of the user. In passive attacks, aggressors do not interrupt the normal performances of the network (Wei, Xiang, Yuebin, & Xiaopeng, 2007). In this attack, the aggressor listens to the network in order to obtain information that what is going on in the network, to know and comprehend how the nodes are exchanging the information with each other, and also how they are situated in the network. Thus before the attacker set up an attack versus the network, the attacker has adequate amount of information about the network that can easily hijack and inject attack in the network.

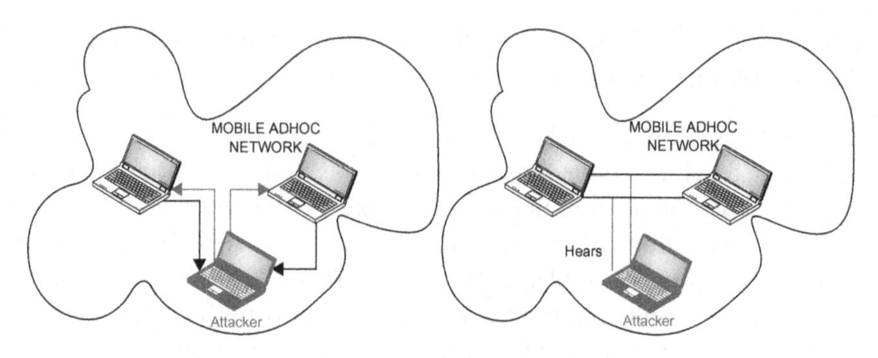


*Figure 2.17 Active and passive attack in MANETs (Ullah & Rehman, 2010).*

In an active attack, the enemy tries to avoid or crack into secured networks. This can be done through turn invisible, malware, malware, or Trojan malware horse. Active attacks include efforts to avoid or crack protection features, to present harmful code, and to grab or change information. In active attack the attacker interrupts the performance of the network, embezzle important information and try to demolish the data during the communication in the network (Wei et al., 2007). Active attacks can be internal or external. The active attacks use to demolishing the performance of network in such a way that active attack operate as internal node in the network. When attacker is an active part of the network, it will be easier for the node to utilize and hijack any internal node to use it to introduce bogus packets injection or DoS. This attack gives the attacker a powerful situation where attacker can alter, mimic, and rebroadcast the messages. Fig. 2.17 displays active and passive attack in MANETs.

3. Inside versus outside attacks (Saha et al., 2012):
   In a core attack, an enemy has affected or taken a node, thus accessing security and verification important factors. The primary method of discovering and mitigating core attacks is to observe the bundle sending actions among the nodes. In an outsider attack, the assailants are believed to have no knowledge of the important factors that are used to secure and verify the information and routing control packages. Avoiding outside assailants from tampering with the information is achieved by simply employing security and verification techniques.
4. Layered attacks (Saha et al., 2012):
   Attacks can also be classified as per the layer at which the attack happens, as indicated in Table 2.1:

**Table 2.1 Layered Attacks (Saha et al., 2012)**

| Layers | Attacks |
|---|---|
| Application layer | Repudiation, data corruption |
| Transport layer | Session hijacking, SYN flooding |
| Network layer | Gray hole, black hole, worm hole, byzantine, Sybil, jellyfish, rushing |
| Link layer | Interception, fabrication, modification |
| Physical layer | Jamming, sniffing |

5. Data versus control traffic attacks (Saha et al., 2012):
   Information traffic attack deals either in nodes losing data packages passing through them or in postponing of sending your details packages. Some types of attacks choose victim packages for losing while some of them drop all of them irrespective of sender nodes. This may highly degrade the service quality and increases end-to-end delay. This also causes significant loss of important data. In management traffic attack, an enemy tries to get accessibility to a valid path by purposely tampering routing messages. In another variation of this attack (eg, replay attack), enemy first listens wireless traffic for control message and then it creates bundle to get accessibility to the path next time when the path request is again sent. In a research done by Jawandhiya, Ghonge, Ali, & Deshpande, (2010), MANET attacks are categorized as following:
6. External vs. internal attacks
   External attackers are chiefly outside the networks who want to gain entrance to the network, and once they gain entrance to the network, they start sending fake packets, DoS in order to interrupt the performance of the whole network. This attack is same as the attacks that are created versus wired network. These attacks can be prevented by implementing security measures such as firewall, where the access of unauthorized person to the network can be mitigated.

Whereas in internal attack the aggressor wants to have normal access to the network and also takes part in the usual activities of the network. The aggressor gain entrance to the network as a new node either by compromising a current node in the network or by malicious impersonation and start its malicious act. Internal attack is more grave attack than external attack. Fig. 2.18 shows external and internal attacks in MANETs.

External attacks (Jawandhiya et al., 2010), in which the enemy is designed to cause blockage, distribute bogus routing information

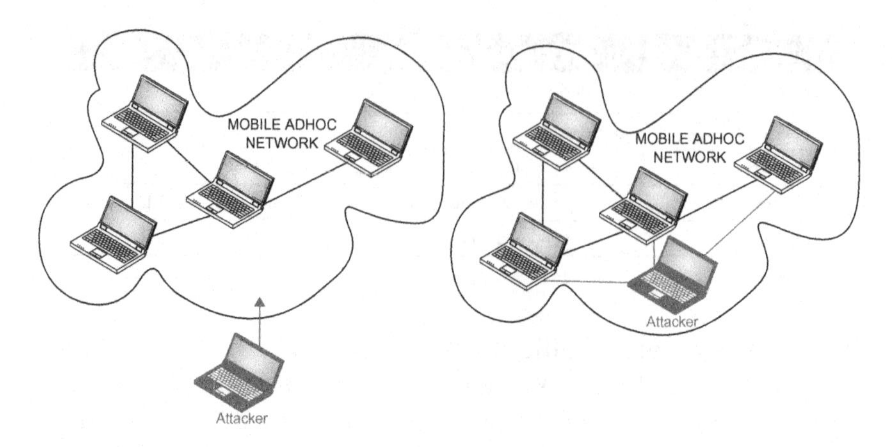


*Figure 2.18 External and internal attacks in MANETs (Ullah & Rehman, 2010).*

or affect nodes from providing services. Internal attacks, in which the enemy wants to, gain the normal accessibility to the network and take part the network activities, either by some harmful impersonation to get the accessibility to the network as a new node or by straight limiting a current node and using it as a basis to perform its harmful actions. Therefore, the security attacks in MANET can be approximately categorized into two major categories (Jawandhiya et al., 2010), namely passive attacks and active attacks are as described in Fig. 2.19. The active attacks further separated according to the levels.

1. Passive Attacks:
   A passive attack (Jawandhiya et al., 2010) does not affect the regular function of the network; attacker snoops the information interchanged in the network without changing it. Here needing privacy gets breached. Recognition of inactive attack is very challenging since the function of the network itself does not get impacted. One of the alternatives to the issue is to use highly effective security procedure to secure the information being passed on, thereby making it challenging for the enemy to get useful information from the information expense. On the other hand, the passive attacks do not affect the network performing (Agrawal et al., 2010).
2. Active attacks:
   An active attack (Jawandhiya et al., 2010) efforts to improving or eliminating the information being interchanged in the network

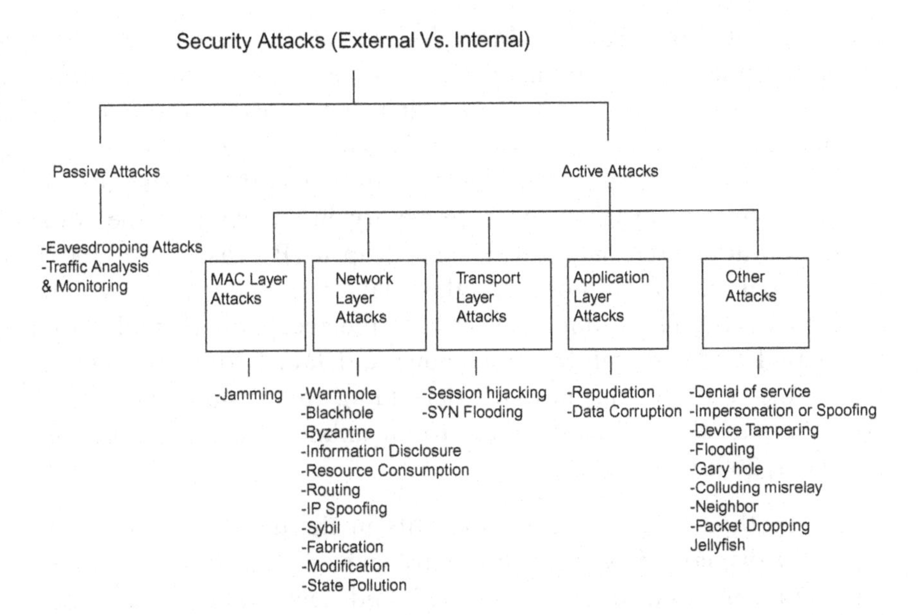


*Figure 2.19 Different types of attacks on MANET (Jawandhiya et al., 2010).*

thereby interfering with the regular performance of the network. Active attacks can be internal or external. External attacks are performed by nodes that are not part of the network. Internal attacks are from affected nodes that are aspect of the network. Since the enemy is already part of the network, internal attacks are more severe and hard to identify than external attacks. Active attacks, whether performed by an exterior advisory or an inner affected node includes actions such as impersonation, modification, production, and duplication.

In active attacks (Agrawal et al., 2010), the regular operation of the network is effective. The enemy has to definitely get involved in the continuous network for interfering with the network performance, hence holds energy and cost to perform the attack. It can eliminate or change the information conveyed in the network. Effectively it degrades the performance or befuddles the routing procedure. The harmful nodes responsible for effective attacks might be due to inner or exterior attacks. The inner attacks are through genuine nodes of the network but are broken or limited against security. They are more difficult to be determined. On the other hand an exterior enemy is an illegal node intruding in the network. They are relatively easy to be protected by means of fire walls, source verification, or security procedure.

## 2.13.2 Black Hole Attack in MANETs

In black hole attack, a malicious node uses its routing protocol in order to publicize itself for having the shortest route to the destination node. This aggressive node publicizes its availability of fresh routes regardless of checking its routing table. In this attack, attacker node always has the accessibility in replying to the route request so adapt the data packet and drop it (Biswas & Ali, 2007). In protocol based on flooding, the malicious node reply will be received by the requesting node before the reception of reply from any actual node; therefore a malicious and faked route will create. When this route set up, now it's depending to the node whether to drop the packets or forward them to an unknown address (Pegueno & Rivera, 2006).

The method how malicious node fits in the data routes changes. Fig. 2.20 displays how black hole problem appears, here node "A" wants to communicate to node "D" and send data packets, and thus the route discovery process initiates. Node "C" is a malicious node then it will claim that it has active route to the specified destination node as soon as it receives RREQ packets from node "A". It will then send RREP to node "A" before any other actual node. In this way, node "A" will assume that this is the active route so active route discovery is completed. After that node "A" will ignore all other replies and start sending data packets to node "C". And finally node "C" will drop all the data packets so they will be consumed or lost.

MANETs face various security threats like attacks that perform against them to interrupt the normal performance of the networks.

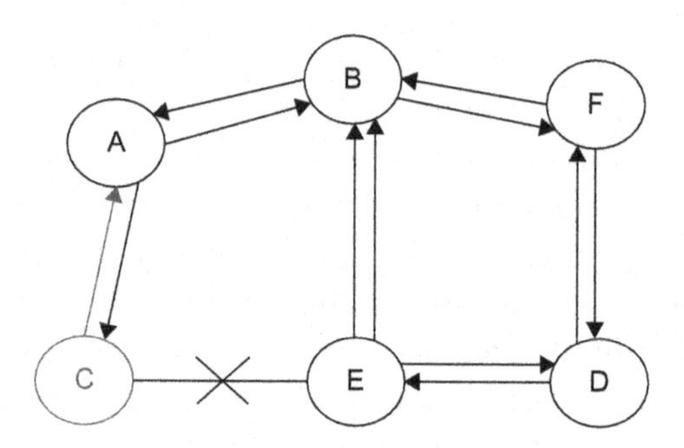


*Figure 2.20 Black hole problem.*

These attacks classified in previous section "Security Issues in MANET" on the basis of their nature. In these attacks, black hole attack is one type of attack which occurs in MANETs. Here, we will describe black hole attack and other attacks that act against MANETs.

### 2.13.3 Black Hole Attack in AODV

There are two types of black hole attack that can be explained in AODV in order to differentiate the kind of black hole attack.

1. Internal black hole attack
   Internal black hole attack has an internal malicious node which fits in between source and destination routes. As soon as this malicious node gets the chance, it makes itself an active data route element. At this point, it is now able to perform attack with initiation of data transmission. Since the node itself belongs to the data route, so this attack is an internal attack. Networks are more vulnerable against internal attack because of trouble in detecting the internal misbehaving node.
2. External black hole attack
   External attacks physically stay outside the network and decline access to network traffic or making congestion in network or by disturbing the whole network. When external attack takes control of internal malicious node and control it to attack to other nodes in MANET, it can become a kind of internal attacks. External black hole attack can be summarized in below steps:
   1. Malicious node discovers the active route and notes the destination address.
   2. An RREP including the destination address field spoofed to an unknown destination address sends by the malicious node. Hop count value is set to lowest values and the sequence number is set to the highest value.
   3. RREP sends to the nearest available node which is part of the active route by the malicious node. This can also send information directly to the source node if route is available.
   4. The RREP received by the nearest accessible node to the malicious node will relay via the set up inverse route to the information of source node.
   5. The source node updates its routing table after receiving new information in the route reply.

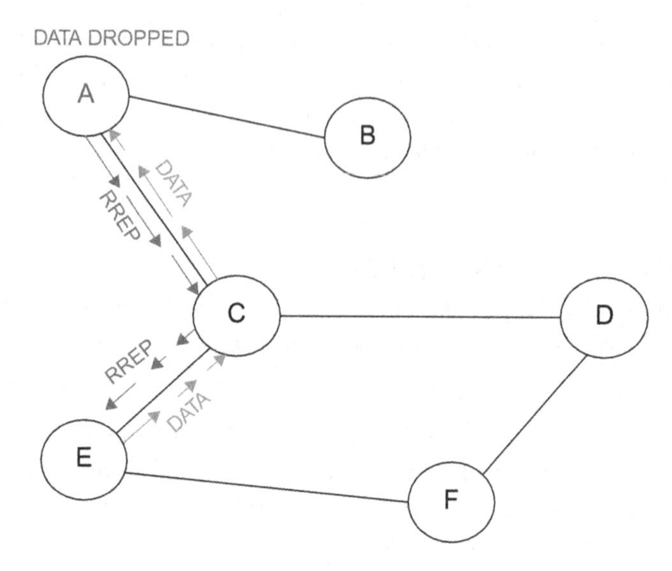


*Figure 2.21 Black hole attack specification.*

6. New route chosen by source node for choosing information.
7. Now, the malicious node will drop all the information to which it belongs in the route.

In AODV black hole attack as you can see in Fig. 2.21, first the malicious node “A” discovers the active route among sender node “E” and destination node “D”. After that, the malicious node “A” sends RREP message which contains the spoofed destination address including small hop count and large sequence number than normal to node “C”. Node “C” forwards RREP to the sender node “E”. Now sender uses this route to send the information and in this way the malicious node will receive the information. And the information will then be dropped. In this way, sender and receiver node will be in no situation any more to broadcasting information in state of black hole attack.

### 2.13.4 Black Hole Attack in OLSR

In OLSR black hole attack, a malicious node forcefully picks up itself as MPR. Malicious node keeps its functionality as continuously in its HELLO message. So in this case, neighbors of malicious node will always select it as MPR. Therefore the malicious node earns a privileged position in the network which utilizes it to perform the DoS attack. This attack will have much more dangerous effect when more than one malicious node appears near the sender and destination nodes.

As research performed by Bhattacharyya et al. (2011) and Saha et al. (2012), black hole attacks can be categorized under information traffic attacks. According to the research of Jawandhiya et al. (2010), Saha et al. (2012), and Bhattacharyya et al. (2011), black hole attack is an effective attack that has been categorized under network part attacks.

The black hole attack has two qualities (Jawandhiya et al., 2010). First, the node uses the mobile ad hoc routing method, such as AODV, to promote itself as having a real path to a location node, even though the path is unwarranted, with the objective of intercepting packages. Second, the enemy takes in the intercepted packages without any sending. However, the enemy operates the risk that nearby nodes will observe and reveal the continuous attacks. There is a more simple way of these attacks when an enemy precisely ahead packages (Jawandhiya et al., 2010). An enemy eliminates or changes packages via some nodes, while making the information from the other nodes unchanged, which boundaries the doubt of its wrongdoing.

In black hole attack (Bhattacharyya et al., 2011; Sharma & Gupta, 2009), a harmful node functions like a black hole, losing all information packages moving through it as like issue and power vanishes from our galaxy in a black hole. If the fighting node is a linking node of two linking elements of that network, then it successfully distinguishes the network in to two turned off elements. Fig. 2.22 shows how the black hole node separates the network into two parts.

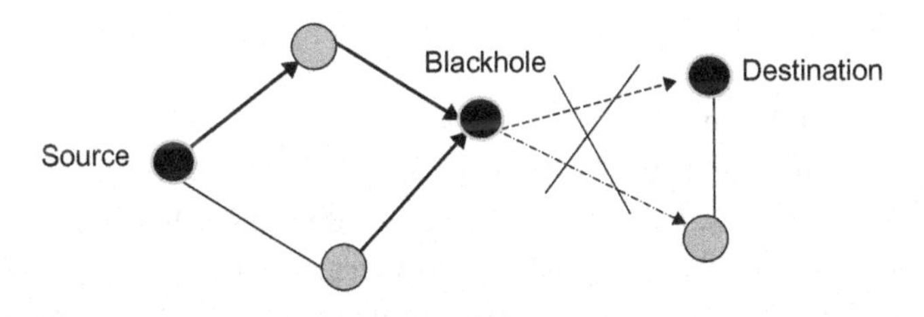


*Figure 2.22 Black hole attack (Bhattacharyya et al., 2011).*

## 2.13.5 Other Attacks in MANETs

### 2.13.5.1 Gray Hole Attack

In this kind of attack the aggressor by accepting to forward the packets in the network misguides the network. Immediately when it receives the packets from the neighbor node, the aggressor drops the packets. This attack is a kind of active attack. At first the aggressor node acts ordinary and replies true RREP messages to the nodes that started RREQ messages. When it receives the packets, it starts dropping the packets and set up DoS attack. The malicious act of gray hole attack in different ways is different. In some gray hole attacks, it drops packets while forwarding them in the network. In other gray hole attacks, the spiteful node acts maliciously for a while until packets drop and then shift to their usual action (Marti, Giuli, Lai & Baker, 2000). Because of this behavior, it is very difficult for the network to comprehend this kind of attack. Gray hole attack is also named as misbehaving attack node (Zhu et al., 2003).

### 2.13.5.2 Flooding Attack

The flooding attack is easy to perform but it brings out the most disturbances. This kind of attack can be attained either by using RREQ or data flooding (Refaei, Srivastava, DaSilva, & Eltoweissy, 2005). In RREQ flooding the attacker floods the RREQ in the whole network that catches a lot of the network sources. The attacker node can be obtained this by selecting IP addresses that do not exist in the network. And thus no node is able to answer RREP packets to these flooded RREQ. The attacker in data flooding gains access into the network and organize routes among all the nodes in the network. Once the routes set up, the attacker injects a boundless amount of futile data packets into the network which go straight to all other nodes in the network. These boundless undesirable data packets overcrowd on the network. Any node that works as a destination node will be occupied all the time by continuously receiving futile and undesirable data.

### 2.13.5.3 Selfish Node

In order to forward packets from one node to another node in MANETs, the nodes execute cooperatively. A selfish node known as a node that decline to work in cooperation to forward packets in order to save its limited resources, this cause chiefly network and traffic interruption (Refaei et al., 2005). A selfish node can decline by publicizing nonexisting paths amid its neighbor nodes or less optimal paths. Saving and preserving the resources is the only concern of selfish node while the network and traffic interruption is the side effect of this behavior. The node can use the network when it needs and after using the network, it turns back to its silent mode. In the silent mode the selfish node is not observable to the network.

The selfish node sometime drops the packets too. If the packets need lots of resources, the selfish node is no longer interested in the packets, so do not forward them in the network and just easily drops the packets.

#### 2.13.5.4 Wormhole Attack

Wormhole attack is a grave attack in which two attackers locate themselves strategically in the network. Then the attackers keep on listening to the network, and record the wireless information. Fig. 2.23 shows the two attackers are located in a strong strategic position in the network.

As we said in wormhole attack, the attackers put themselves in powerful strategic place in the network. They utilize their location, that is, they have shortest route among the nodes as displayed in Fig. 4.3. They publicize their route to other nodes in the network to announce them they have the shortest route for transferring their information. In order to recording the ongoing communication and traffic at one network position and channels them to another position in the network the wormhole attackers create a tunnel (Shanthi et al., 2009).When the attacker nodes create a direct link between each other in the network, then the wormhole attacker at one side receives packets and transfers them to the other side of the network. When the

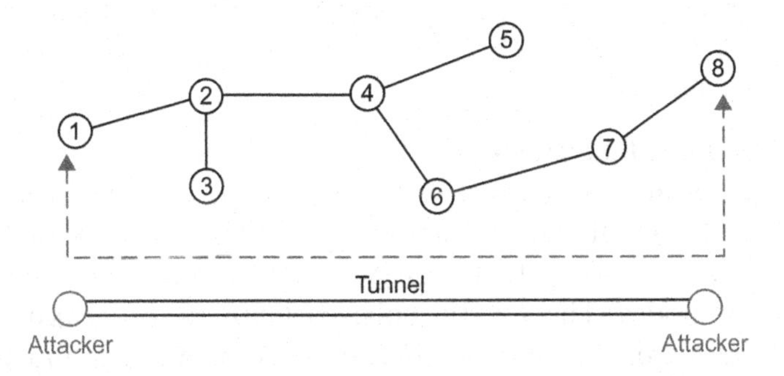


*Figure 2.23 Wormhole attack.*

attackers are in this situation the attack is known as out of band wormhole (Mahajan et al., 2008).

In other type of wormhole attack that is known as in band wormhole attack (Mahajan et al., 2008) the attacker builds an overlay tunnel over the existing wireless medium. This attack is potentially very much dangerous and attacker most prefers to choose this one.

#### 2.13.5.5 Sleep Deprivation Torture Attack

In one of the interesting attacks in MANET the attacker attempts to keep the node awake until all its energy lost and the node go into constant sleep. This attack is known as sleep deprivation torture attack (Stajano & Anderson, 2002). The operating nodes in MANETs have limited resources, that is, battery life, the node stay active for transferring packets during a communication. When the communication stops these nodes go back to sleep mode so that maintain their resources. The attacker makes the node busy, keep it awake so all its energy waste and thus the node goes to sleep for the rest of its life. When the node went to sleep forever, an attacker can simply go around into the network and utilize remainder of the network.

#### 2.13.5.6 Jellyfish Attack

The attacker in jellyfish attack attacks in the network and present undesirable delays in the network (Nguyen & Nguyen, 2008). The attack works as below: first the attacker node gains entrance to the network, once it got into the network and became a part of the network, then by delaying all the packets that receives, it presents delays in the network. When delays are extended, then packets release in the network. This enables the attacker to produce high end-to-end delay, high delay jitter, and significantly impress the performance of the network.

#### 2.13.5.7 Modification Attack

The nature of ad hoc network is that any node can join freely to the network and leave it. Spiteful nodes which plan to attack first join to the network. And then it utilizes lack of symmetries in the network between the nodes. The spiteful node participates in the broadcasting process afterwards on some point establishes the message modification attack (Wei et al., 2007). Misrouting and impersonation attacks are two types of modification attack.

#### 2.13.5.8 Misrouting Attack

A spiteful node in misrouting attack which is part of the network, attempts to reroute the traffic from its starting node to an erroneous and strange destination node. As long as the packets abide in the network keep using of network resources. And when the packet cannot find its destination the network drops the packet.

#### 2.13.5.9 Impersonation Attack

In ad hoc networks nodes can move freely inside and outside of the network. There is not any secure authentication process in order that make the network secure from spiteful nodes. In MANETs the host has been identified by IP and MAC address uniquely. But these measurements are not sufficient to authenticate sender. For obtaining identity of another node and hide into the network the attacker uses MAC and IP spoofing, so this attack is also known as spoofing attack (Wei et al., 2007).

#### 2.13.5.10 Routing Table Overflow Attack

Routing Table Overflow attack usually acts versus proactive protocols. In this attack, no-existent node information send in the network, furthermore corrupting and degrading the speed, when routing tables are updated. Proactive routing protocols updates route periodically before even they are required. This is one of the weaknesses that make proactive protocols vulnerable to the routing table attack. The attacker attempts to make so many routes to nodes that do not exist in the network. They do it by using RREQ messages. The attacker sends RREQ messages in the network to nonexistent nodes. Routing tables of the nodes under attack get full and does not have any more entry to make new. In other words the routing tables of the attacked nodes overflow with so many route entries (Deng et al., 2002).

## 2.14 RELATED STUDIES

Table 2.2 presents some of the related studies on existing solutions for multiple black hole attack or in similar domains. These studies were found useful and have provided a focus for this study. The table is based on the following headings:

1. Title of the study.
2. Names of author.
3. Briefly description of the studies.
4. Experimental results.
5. Study limitations.

**Table 2.2 Related Studies**

| Title of the Study | Author | Brief Description of the Study | Experimental Results | Study Limitations |
|---|---|---|---|---|
| An efficient prevention of black hole problem in AODV routing protocol in MANET | (Singh & Sharma, 2012) | The proposed method uses promiscuous mode of a node to overhear the neighbor's communication. And propagates the information of malicious node to all the other nodes in the network. | The experimental results show minimum routing overhead to combat the black hole problem. It does not require any database, extra memory, and more processing power. | Nil |
| Preventing black hole attack in MANETs using randomized multipath routing algorithm | (Vincent & Meshach, 2012) | This chapter contributes a mechanism that generates random multipath route which is circumventing black holes. | Experimental results show that based on the proposed routing metric (delay of transmission packets, the bandwidth of the channel, throughput of the packet transmission) is introduced. During path maintenance, predicted signal strength and channel average fading duration are combined with handoff strategy to combat channel fading and improve channel utilization. The packet loss is completely overcome. | This technique does not provide any security against the black hole attack which is the commonly known attack on MANETs. |
| Improving AODV protocol against black hole attacks | (Mistry et al., 2010) | The solution that is proposed here is designed to prevent any alterations in the default operations of either the intermediate nodes or that of the destination nodes. The approach we follow, basically only modifies the working of the source node, using an additional function. | Experimental results show significant improvement in packet delivery ratio of AODV in presence of black hole attacks, with marginal rise in average end-to-end delay. | Nil |
| DPRAODV: a dynamic learning system against black hole attack in AODV-based MANET | (Raj & Swadas, 2009) | In this chapter, it proposed a DPRAODV (detection, prevention and reactive AODV) to prevent security threats of black hole by notifying other nodes in the network of the incident. | Experimental results show that the prevention scheme detects the malicious nodes and isolates it from the active data forwarding and routing and reacts by sending ALARM packet to its neighbors. The solution: DPRAODV increases PDR with minimum increase in average-end-to-end delay and normalized routing overhead. | Nil |

| | | | | |
|---|---|---|---|---|
| Routing security in wireless ad hoc networks | (Deng et al., 2002) | The proposed solution uses one more route to the intermediate node that replays the RREQ message to check whether the route from the intermediate node to the destination node exists or not. If it exists, we can trust the intermediate node and send out the data packets. If not, we just discard the reply message from the intermediate node and send out alarm message to the network and isolate the node from the network. | Feasible solution for [illegible] AODV protocol. | proposed method is that it works based on an assumption that malicious nodes do not work as a group, although this may happen in a real situation. |
| AOMDV routing-based enhanced security for black hole attack in MANETs | (Geetha & Revathi, 2013) | The idea behind multipath routing is to look for a multiple routes to a host with the intention of avoiding black hole attack. There could be a lot of reasons to do this, if the black hole attack occurs in a single path, the AOMDV will send the data packets in some other route which is available in the multipath routing. | Experimental results show: compared to the existing AODV protocol, AOMDV has better packet delivery ratio and comparatively low average end-to-end delay. The number of packets dropped in the AOMDV against the black hole attack is very low. Thus the proposed technique which uses AOMDV is proved to be better against black hole attacks. | Nil |
| Performance evaluation on modified AODV protocols | (Ahmad et al., 2012) | The purpose of this chapter is to evaluate some of the modified AODV protocols performance by examining their effectiveness in alleviating the black hole attack and further examining the effect of mitigation methods used on overhead. The performance analysis focuses on two conditions, that is, no attack and under attack. Three modified AODV protocols were studied, namely IDSAODV, HDAODV and EAODV, and a new modified protocol is proposed. | The results showed that the three modified AODV protocols give positive effect to network performance whether network is under-attack or no-attack in comparison with normal AODV performance. EAODV have the potential of becoming a preferred protocol to mitigate black hole problem, but we need to address longer delay and higher energy usage. | Nil |
| Simulation of black hole attack in wireless ad hoc networks | (Dokurer, 2006) | In this study, the black hole attack in wireless ad hoc networks and evaluated its damage in the network has simulated using NS-2 simulation program. Has implemented a new routing protocol which simulates the black hole. Then, it proposed an IDS solution to eliminate the black hole effects in the AODV network. The solution has implemented into the NS-2 also. At the end, it evaluated the results. | As a result, the proposed solution is eliminated the black hole effect with 24.38% success. | Nil |

## 2.15 INVESTIGATED SOLUTIONS

The selected solutions for this project which carried out after investigating existing solutions for preventing multiple black hole attacks in MANET using AODV routing protocol are as following:

1. DPRAODV method, presented by Raj and Swadas (2009).
2. AOMDV method, presented by Geetha and Revathi (2013).
3. Modified AODV, presented by Mistry, Jinwala, and Zaveri (2010).
4. IDSAODV method, presented by Dokurer (2006).

## 2.16 INTRUSION DETECTION SYSTEM (IDSAODV)

(Dokurer, 2006) proposed IDSAODV, which is another modified AODV that is designed to reduce the adverse effect of black hole attack. The protocol mitigation method is implemented by modifying the routing update mechanism in AODV protocol. The process to ignore the first establishment route is added to the logical expression in routing update process. The main strategy is that when the network is under attack, multiple RREP from a different path is generated. This protocol assumes that the first RREP message that arrived at a node is from a malicious node, and hence the mitigation method in IDSAODV is to ignore this RREP to avoid false route entry being updated to the routing table. This method is able to improve the packet delivery but there is at least one limitation, for example, if the second RREP message received at a source node comes from a malicious node, it is not able to avoid or stop it.

## 2.17 EVALUATION METRICS

The metrics used to evaluate the performance are given below:

Packet delivery ratio: The ratio between the client of packages originated by the "application layer" CBR sources and the number of packages received by the CBR sink at the final destination:

$$Packet\ Delivery\ Ratio = \frac{\sum Number\ of\ packet\ receive}{\sum Number\ of\ packet\ send} \tag{2.1}$$

Average end-to-end delay: This is the common wait between the delivering of the information bundle by the CBR resource and its invoice at the corresponding CBR recipient. This contains all the setbacks triggered during path purchase, streaming and handling at advanced nodes, retransmission setbacks at the MAC part, etc.

Average end-to-end delay of the application data packets, denoted by D, is calculated as follows:

$$D = \frac{\sum_{i=1}^{n} d_i}{n} \quad (2.2)$$

Where $d_i$ is the average end-to-end delay of data packets of $i^{th}$ application and $n$ is the number of CBR applications.

Routing overhead: This concept is the ratio of number of control packet generated to the data packets transmitted.

Control packet overhead is the ratio of the number of control data bytes which is used by the sender to discover the secure route between sender and receiver and the total number of application data bytes transferred between sender and receiver. This is denoted by $O$ and calculated as follows:

$$O = \frac{\left(\sum_{i=1}^{n} rrq_i + \sum_{i=1}^{n} rrt_i\right) \times cpsize}{\sum_{i=1}^{n}(N_i^s \times dpsize_i) + \left(\sum_{i=1}^{n} rrq_i + \sum_{i=1}^{n} rrt_i\right) \times cpsize} \times 100\% \quad (2.3)$$

The $rrq_i$ is the number of route requests sent by the sender and $rrt_i$ is the number of route request retries done by the sender. *cpsize* is the size of the request packet in bytes, and $dpsize_i$ is the size of the application data packet in the $i^{th}$ application. $N_i^s$ is the number of data packets sent by the $i^{th}$ application source and $n$ is the number of applications.

## 2.18 SUMMARY

This chapter covered important aspects of MANET. This chapter also included concepts depending on routing and routing protocols in MANET, attacks which effect on these routing protocols, more explanation about AODV and its vulnerabilities. After that, we will review related studies about black hole attack and will go over the solutions that already presented for preventing of this attack. And, at the end, it explained some evaluation metrics which use for network performance investigation.

# CHAPTER 3

# Research Methodology

## 3.1 INTRODUCTION

This chapter describes the research methodology of the project in classifying the systematic work achieved through a series of steps used as a guideline for achieving the objectives of this research. This study is based on comparison between effects of efficient method for solving single black hole attack and its effects on multiple black hole attacks in mobile ad hoc network (MANET) using AODV routing protocol. The study investigates the effectiveness of the selected solution based on the following metrics; packet delivery ratio, packet loss percentage, average end-to-end delay, and route request overhead. Therefore, this chapter will clearly discuss the guideline to achieving the research goals and objectives of this study.

## 3.2 RESEARCH STRUCTURE

Research framework is used to diagrammatically describe the individual steps followed throughout this research. In general, it is used as a guide by researchers to zoom-in on the scope of study. Fig. 3.1 shows the research framework followed in this study. The study is divided into three phases and each phase's output is an input to the next phase. Phase 1 is based on the investigation of the existing solutions using AODV routing protocol for preventing single black hole attacks in MANET. Phase 2 is concluded with two phases: Phase 2a is focused on determining one efficient solution between the existing solutions from phase 1 that implemented by NS2 before based on the following metrics: packet delivery ratio, packet loss percentage, average end-to-end delay, and route request overhead. Phase 2b implements the proposed solution on MANET without black hole attack and MANET with two black hole attacks. Finally, phase 3 compares the results found out from phase 2 in finding the performance of single black hole attack prevention solutions on network with multiple black hole attacks. The phases are indicated in Fig. 3.1.

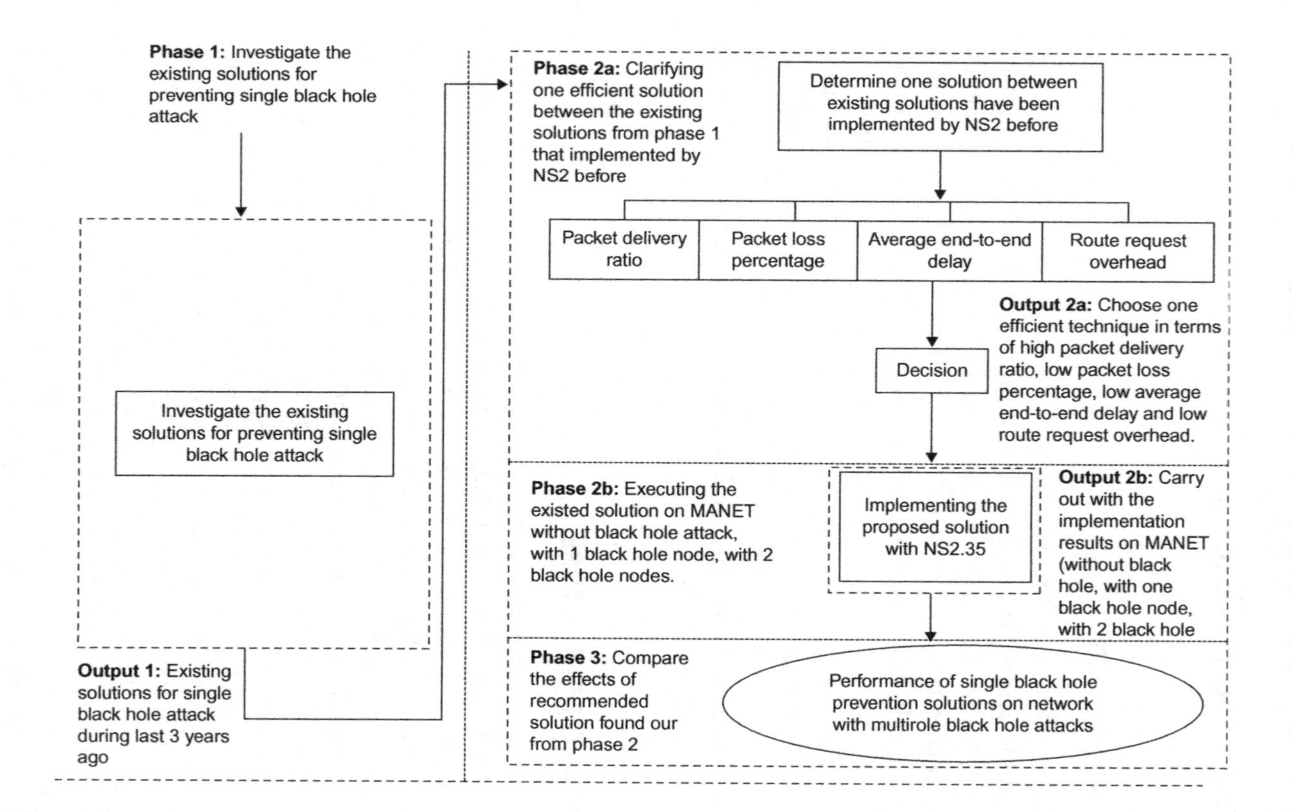


*Figure 3.1 Research framework.*

The research will be conducted through four main phases. The following subsections will describe each phase briefly.

### 3.2.1 Phase 1: Investigating the Existing Solutions

Investigating the information in this phase has been done as following:

1. Searching has been done between existing solutions since 3 years ago.
2. Information is collected from http://ieeexplore.ieee.org/, http://www.sciencedirect.com/ and http://scholar.google.com.my/.
3. For choosing the proposed solutions in this phase, it is looked for simple and efficient methods.

### 3.2.2 Phase 2

#### 3.2.2.1 Phase 2a: Clarifying Efficient Solution

In phase 1, the solutions that investigated will be used in this phase to determine a good solution in terms of high packet delivery ratio, low end-to-end delay, low packet loss percentage, and lessened route request overhead. In this phase, we determined the proposed solution between solutions that already have been implemented by NS2.

#### 3.2.2.2 Phase 2b: Executing the Existed Solution

In this phase the solution that is determined in phase 2a will be implemented. For implementation, NS-2.35 simulator will be used and implementation will be done on MANET without black hole attack, MANET with one black hole node, and MANET with 2 black hole nodes.

##### 3.2.2.2.1 Simulation: The Customary Definition

According to Issariyakul (2012), simulation is "the process of designing a model of a real system and conducting experiments with this model for the purpose of understanding the behavior of the system and/or evaluating various strategies for the operation of the system." With the dynamic nature of computer networks, we thus actually deal with a dynamic model of a real dynamic system.

##### 3.2.2.2.2 Network Simulator (NS)

In this study, an event-driven network simulator tool known as NS2 (version 2.35) is used, and according to Issariyakul (2012), it has been proved efficiently capable in the study of dynamic nature of communication networks. Furthermore, wired and wireless network-related protocol can be simulated using NS2. Practically, NS2 provides a way for users to specify network protocols in the field or wired and wireless network and also simulating their individual behaviors based on predefined parameters.

Furthermore, NS2's publicity and the variations of different versions revised by different institutions after its original development at University of California and Cornell University in 1989. Institutional unit such as Defense Advanced Research Projects Agency (DARPA) took interest in the development of NS via the Virtual InterNetwork Testbed (VINT) project (Issariyakul, 2012). National Science Foundation (NSF) among others have joined in the continuous revision and development of NS2 including group of community researchers and developers who are continuously working to make NS2 better. The simulation is done to analyze the performance of MANET without black hole attack, with 1 black hole mobility node, 2 black hole mobility nodes, and 3 black hole mobility nodes in terms of below metrics:

1. Packet delivery ratio.
2. End-to-end delay.
3. Packet loss percentage.
4. Route request overhead.

NS is an event-driven network simulator program, developed at the University of California, Berkley. It includes several network objects such as protocols, applications, and traffic source behavior. The NS is a part of VINT project being supported by DARPA since 1995 (Issariyakul & Hossain, 2012).

The NS-2 at the simulation layer, interprets user simulation scripts by using OTcl (object-oriented tool command language) programming language. OTcl language is an object-oriented extension of the Tcl Language which is fully compatible with the C++ programming language. At the top layer, NS is an interpreter of users' Tcl scripts; both make use of C++ codes. Further usage of the Tcl will be explained in-depth in Chapter 4.

Fig. 3.2 shows the use of NS-2 in interpreting OTcl script. Furthermore, while OTcl script is being interpreted, NS creates two main analysis reports simultaneously. One of the reports created by NS is known as NAM (Network Animator) object that shows the visual animation of the simulation. The second report is the trace

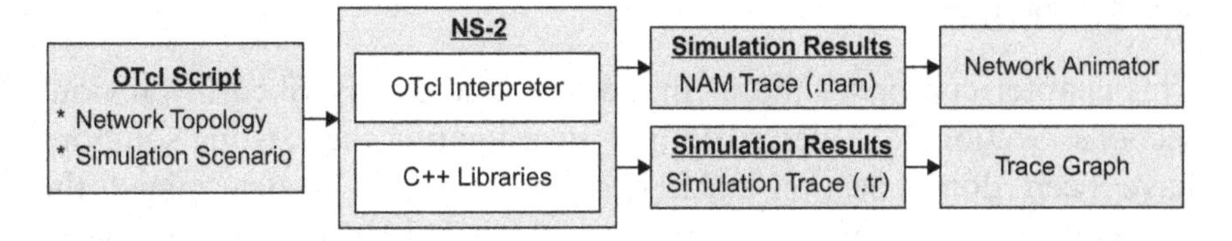


*Figure 3.2 NS-2 schema (Dokurer, 2006).*

object that describes the behavior of all objects in the simulation. Both of them are created as a file by NS. Former is .nam file implemented in NAM software that comes along with NS. Latter is a ".tr" file that includes all simulation traces in the text format.

Although, NS project is normally distributed inclusive of various packages (ns, nam, tcl, otcl etc.) known as "all-in-one package," they can also be downloaded separately. In this study, NS-2 version 2.35 of "all-in-one package" was installed in Windows environment using Cygwin. After version 2, NS is commonly using NS-2, and in this book, it shall be referred to as NS-2. The ".tcl" files was written in text editor and the results of the ".tr" file have been analyzed using "cat," "awk," "wc," and "grep" commands in Unix operating system. The implementation phase of the black hole behavior to the AODV protocol is written using C++.

3.2.2.2.3 Tool Command Language (Tcl) in NS

Tool Command Language (Tcl) is a powerful interpreted programming language developed by John Ousterhout at the University of California, Berkeley (Bessire, 2006). Tcl is a very powerful and dynamic programming language. It has a wide range of usage, including web and desktop applications, networking, administration, testing, etc. Tcl is a truly cross platform, easily deployed and highly extensible. The most significant advantage of Tcl is that it is fully compatible with the C programming language and Tcl libraries can be interoperated directly into C programs. We shall describe the Tcl code and we have designed to implement the black hole attacks in the next section.

### 3.2.3 Phase 3: Comparing the Effects of Recommended Solution on MANET Performance

This phase will compare the results carried out in previous phase to realize the effects of single black hole attack solutions on MANET performance with multiple black hole attacks.

## 3.3 SUMMARY

This chapter comprises the methodology used as described in previous sections. Section 3.2.1 described how investigating the existing solutions have been done. Section 3.2.2 described how it is determined the efficient solution using different metrics (packet delivery ratio, packet loss percentage, average end-to-end delay, and route request overhead). Finally, Section 3.2.3 described the comparison of proposed solution effects on MANET performance.

CHAPTER 4

# Investigation and Selection Procedure

## 4.1 INTRODUCTION

This chapter discusses the effects of mobile ad hoc black hole attacks in the networks. To achieve this, it simulated the mobile ad hoc network scenarios which include black hole node using NS Network Simulator program (Issariyakul & Hossain, 2012). To simulate the black hole node in a mobile ad hoc network, this project implemented a new protocol that drops data packets after receiving them. In this section, having shown how it's tested with the black hole implementation, it will present the simulations of black hole attack to demonstrate its effects. Then, it will evaluate the effects of black hole attacks on a mobile ad hoc network.

## 4.2 EXECUTING A NEW ROUTING PROTOCOL IN NS TO SIMULATE BLACK HOLE BEHAVIOR

Ros and Ruiz (2004) described the implementation of a new MANET unicast routing protocol in NS-2. The initial implementation of this project is based on the work of Ros and Ruiz (2004). In this research, nodes that exhibit black hole behavior in mobile ad hoc network using AODV routing protocol is used. Since the nodes behave as a black hole, they have to use a new routing protocol that can participate in the AODV messaging. Implementation of this new routing protocol is explained below in detail:

All routing protocols in NS are installed in the directory of "ns-2.35." First and foremost, the AODV protocol found in this directory is being duplicated after which the directory is renamed as "blackholeaodv". Names of all associated aodv files labeled as "aodv" in the directory are changed to "blackholeaodv," that is, *blackholeaodv.cc*, *blackholeaodv.h*, *blackholeaodv.tcl*, *blackholeaodv_rqueue.cc*, *blackholeaodv_rqueue.h*, etc. in this new directory excluding "*aodv_packet.h*."

The main concept is such that AODV and black hole AODV protocols send the same AODV packets. Hence, we do not copy "*aodv_packet.h*" file into the blackholeaodv directory.

Changes have been made to all classes, functions, structs (except struct names that belong to AODV packet.h code), variables, and constants names in all the files in the directory. Also, the aodv and blackholeaodv protocols have been designed to send each other aodv packets. These two protocols are actually the same.

Following the above changes, two common files that are used in NS-2 globally to integrate new blackholeaodv protocol to the simulator have also been changed. In the work of Ros and Ruiz (2004), more files are changed to add new routing protocol and this new protocol uses its own packets. But in this research work, we do not need to add a new packet. Therefore, only two files were changed. The changes are explained below.

The First file modified is "\tcl\lib\ns-lib.tcl" such that protocol agents are implemented as a procedure. In the case of the nodes using blackholeaodv protocol, this agent is scheduled at the beginning of the simulation and it is designated to the nodes that will use blackholeaodv protocol. The agent procedure for blackholeaodv is shown in Fig. 4.1:

```
switch -exact $routingAgent_ {
     DSDV {
          set ragent [$self create-dsdv-agent $node]
     }
     DSR {
          $self at 0.0 "$node start-dsr"
     }
     AODV {
          set ragent [$self create-aodv-agent $node]
     }
     blackholeAODV {
          set ragent [$self create-blackholeaodv-agent $node]
     }
Simulator instproc create-blackholeaodv-agent { node } {
      set ragent [new Agent/blackholeAODV [$node node-addr]]
      $self at 0.0 "$ragent start" :   # start BEACON/HELLO Messages
      $node set ragent_ $ragent
      return $ragent
}
```

*Figure 4.1 "blackholeaodv" protocol agent is added in "\tcl\lib\ns-lib.tcl."*

Second file which modified is "*\makefile*" in the root directory of the "ns-2.35." After all implementations are ready, we have to compile NS-2 again to create object files. We have added the below lines in Fig. 4.2 to the "*\makefile.*"

```
blackholeaodv/blackholeaodv_logs.o blackholeaodv/blackholeaodv.o \
blackholeaodv/blackholeaodv_rtable.o blackholeaodv/blackholeaodv_rqueue.o \
```

*Figure 4.2 Addition to the "\makefile."*

To this extent, implementation of a new routing protocol labelled as blackholeaodv have been done. But black hole behaviors have not yet been implemented in this new routing protocol. To include black hole behavior into the new AODV protocol, same changes were made in blackholeaodv/blackholeaodv.cc C++ file. Description of these changes made in blackholeaodv/blackholeaodv.cc file explaining working mechanism of the AODV and black hole AODV protocols are as below.

In a scenario where a packet is received by the "*recv*" function of the "*aodv/aodv.cc*," the packets are being processed based on its type. In case the type of packet is associated with any of AODV route management packets, it sends the packet to the "*recvAODV*" function that will be explained below. If the received packet is a data packet, typically AODV protocol sends it to the destination address; on the other hand, behaving as a black hole, it drops all data packets as long as the packet does not come to itself. In the code below, the first "*if*" condition provides the node to receive data packets if it is the destination. The "*else*" condition drops all remaining packets. The "*if*" statement is shown in Fig. 4.3:

```
//If destination address is itsself
if ( (u_int32_t)ih->saddr() == index)
    forward((blackholeaodv_rt_entry*) 0, p, NO_DELAY);
else
// For blackhole attack in the wireless adhoc network, after taking
the path over itself, misbehaving node
    drop(p, DROP_RTR_ROUTE_LOOP);
```

*Figure 4.3 "If" statement for dropping or accepting the packets.*

If the packet is associated AODV management packet, "recv" function sends it to "recvblackholeAODV" function. The "recvblackholeAODV" function checks the type of the AODV management packet and based on the packet type it sends them to appropriate function with a "case" statement. For instance, RREQ packets are sent to the "recvRequest" function, RREP packets to "recvReply" function, etc., and case statements of "recvblackholeAODV" function are shown in Fig. 4.4:

```
 * Incoming Packets.
 */
switch(ah->ah_type) {

case AODVTYPE_RREQ:
  recvRequest(P);
  break;

case AODVTYPE_RREP:
  recvReply(P);
  break;

case AODVTYPE_RERR:
  recvError(P);
  break;

case AODVTYPE_HELLO:
  recvHello(P);
  break;

default:
  fprintf(stderr, "Invalid blackholeAODV type (%x)\n", ah->ah_type);
  exit(1);

}
```

*Figure 4.4 Case statements for choosing the AODV control message types.*

In our case, we will consider the RREQ function because black hole behavior is carried out as the malicious node receives an RREQ packet. When the malicious node receives an RREQ packet, it immediately sends RREP packet as if it has fresh enough path to the destination. Malicious node tries to deceive nodes sending such an RREP packet. Highest sequence number of AODV protocol is 4294967295, 32-bit unsigned integer value. Values of RREP packet that malicious node will send are described below. The sequence number is set to 4294967295 and hop count is set to the false RREP message of the black hole attack is shown in Fig. 4.5:

```
    sendReply (rq->rq_src,                // IP Destination
               1,                         // Hop Count
               index,                     // Dest IP Address
               MY_ROUTE_TIMEOUT,          // Highest Dest Sequence Num
               rq->rq_timestamp);         // Lifetime
                                          // timestamp
    Packet::free(P);
}
```

*Figure 4.5 False RREP message of black hole attack.*

After all changes are finished, we have to recompile all NS-2 files to create object files. For recompiling, we should run make clean, then make depend, and finally make command in ns2.35 directory. Now, we have a new test bed to simulate black hole attack in AODV protocol. The next section will explain the simulations and simulation results.

## 4.3 EXAMINING THE BLACK HOLE AODV

The implementation of the black hole is tested to see whether it is correctly working or not. To be sure the implementation is correctly working, we used the NAM (network animator) application of NS. To test the implementation, we used two simulations. In the first scenario, we did not use any black hole AODV node (the malicious node that exhibits the black hole attack will be called "black hole node"). In the second scenario, we added a black hole AODV node to the simulation. Then, we compared the results of the simulations using NAM.

### 4.3.1 Simulation Parameters and Measured Metrics

To take accurate results from the simulations, we used UDP protocol. The source node keeps on sending out UDP packets, even if the malicious node drops them, while the node finishes the connection if it uses TCP protocol. Therefore, we could observe the connection flow between sending node and receiving node during the simulation. Furthermore, we were able to count separately the sent and received packets since the UDP connection is not lost during the simulation. If we had used TCP protocol in our scenarios, we could not count the sent or received packets since the node that starts the TCP connection will finish the connection after a while if it has not received the TCP ACK packet.

It has generated 10 networks: 6, 20, 30, 40, 50, 60, 70, 80, 90, and 100 nodes, and then creates a UDP connection between node 0 and node 4 in all of scenarios and attach CBR (constant bit rate) application that generates constant packets through the UDP connection. CBR packet size is 512 bytes; data rate is set to 10 Kbyte/sec. Duration of the scenarios is 500 s and the CBR connections started at time 2.0 s and continue until the end of the simulation, in a $670 \times 670$-m flat space. The appropriate positions of the nodes are manually designed to show the data flow and also introduce movement for nodes 0–5 in network with 6 nodes and for nodes 0–9 for networks with 20 and for nodes 0–29 with 30 nodes that can observe the changes of the data flow in networks with different number of nodes. The Tcl script contains a black hole AODV for the first simulation is shown in Appendix A.

### 4.3.2 Appraisal of the Simulation

In the first scenario where there is not a black hole AODV node, connection between node 0 and node 4 is correctly flowed when look at the animation of the simulation, using nam.file. Figs. 4.6–4.11 show the data flow from node 0 to node 4 in three different networks with 6, 20, and 30 mobile nodes. Fig. 4.6 shows data flow for network with six nodes, the connection between node 0 and node 4 established via node 1.

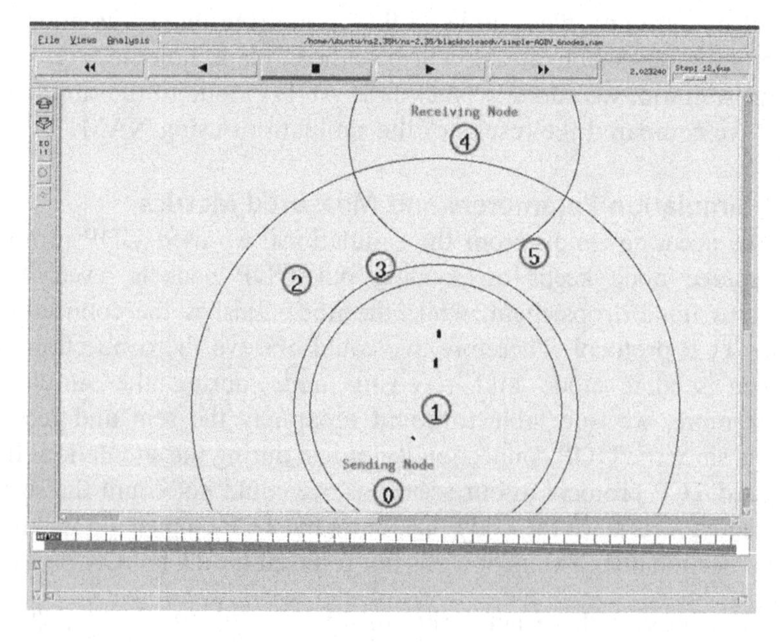


*Figure 4.6 Data flow between node 0 and node 4 via node 1 with 6 mobility nodes.*

In the first network when node 1 leaves the propagation range of the node 0 while moving, the new connection is established via node 3. The new connection path is shown in Fig. 4.7:

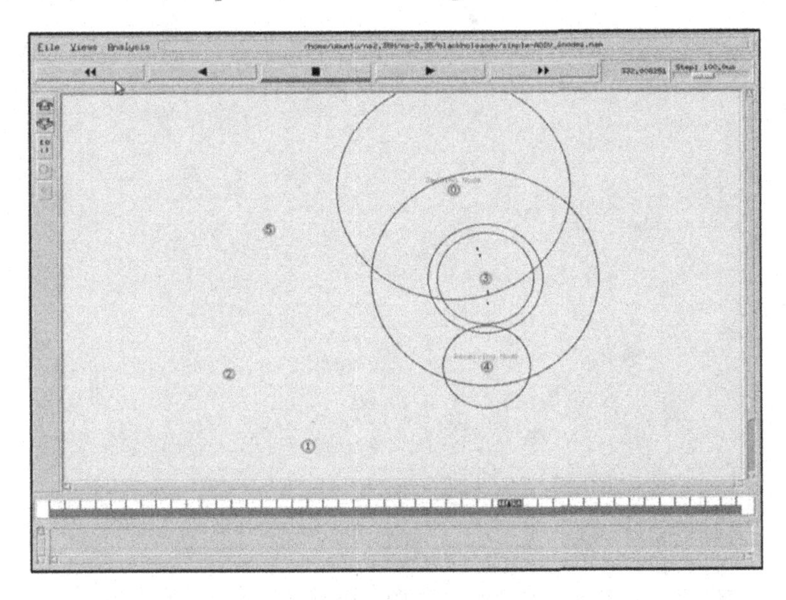

*Figure 4.7 Data flow between node 0 and node 4 via node 3 after mobility.*

Fig. 4.8 shows data flow for network with 20 nodes, the connection between node 0 and node 4 established via nodes 13, 12, and 6:

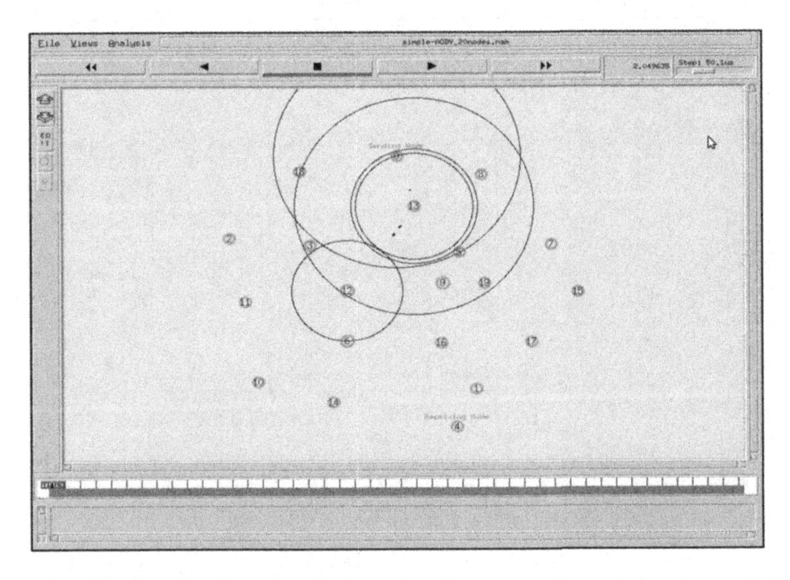

*Figure 4.8 Data flow between node 0 and node 4 via nodes 13, 12, and 6 with 20 mobility nodes.*

In the second network when node 13 leaves the propagation range of the node after moving, the new connection is established via nodes 8, 17, 11, and 2. The new connection path is shown in Fig. 4.9:

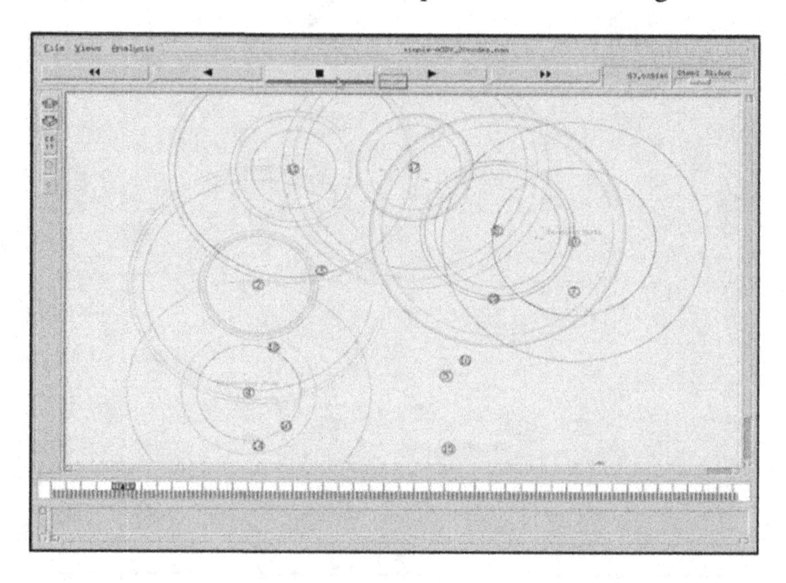

*Figure 4.9 Data flow between node 0 and node 4 via nodes 8, 17, 11, and 2 after mobility.*

Fig. 4.10 shows data flow for network with 30 nodes, that connection between node 0 and node 4 established via nodes 18, 2, and 12:

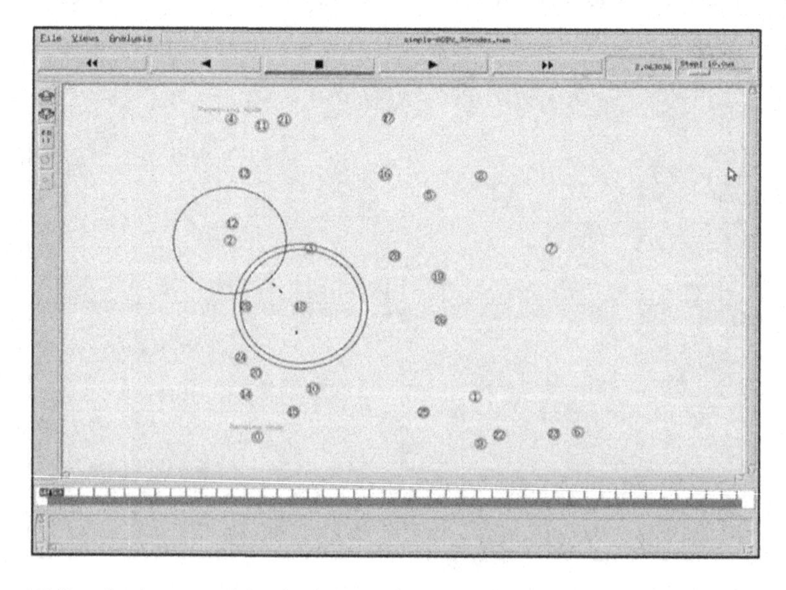

*Figure 4.10 Data flow between node 0 and node 4 via nodes 18, 2, and 12 with 30 mobility nodes.*

In the third network when node 18 leaves the propagation range of the node after moving, the new connection is established via nodes 5, 28, and 10. The new connection path is shown in Fig. 4.11.

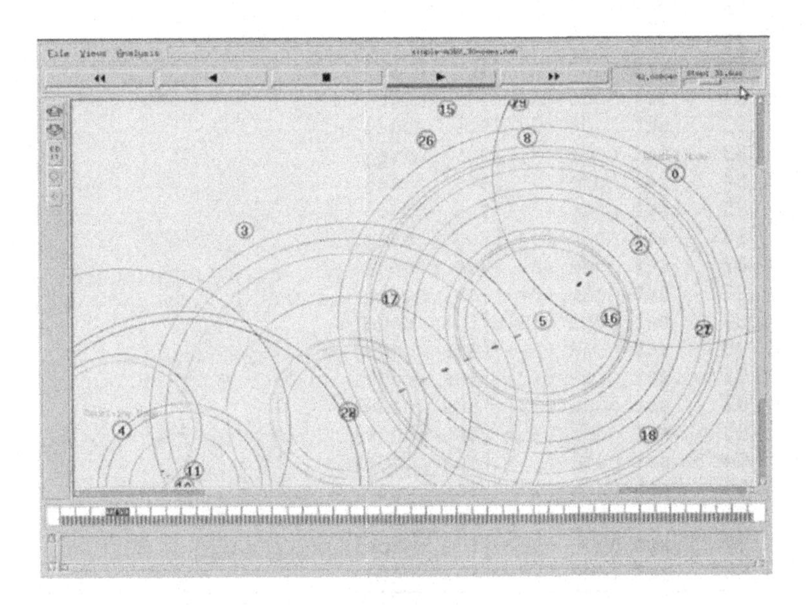


*Figure 4.11 Data flow between node 0 and node 4 via nodes 5, 28, and 10 after mobility.*

## 4.4 SIMULATION OF BLACK HOLE ATTACK

### 4.4.1 Simulation Parameters and Measured Metrics

Different scenarios are used in our work: 6, 20, 30, 40, 50, 60, 70, 80, 90, and 100 AODV nodes without black hole attack; 6, 20, 30, 40, 50, 60, 70, 80, 90, and 100 AODV nodes with one black hole node; and 6, 20, 30, 40, 50, 60, 70, 80, 90, and 100 AODV nodes with two black hole nodes. UDP connections are established between nodes. In all of the scenarios, the sending node is node 0 and the receiving node is node 4 and packets send from sending node to receiving node.

Node positions and movements are generated manually, during 500 s, in a 670 × 670-m space. The CBR application is attached that generates constant packets through the UDP connection. Duration of the scenarios is 500 s and the CBR connections started at time 2.0 s until end of scenario. In our scenarios, CBR parameters are as follows:

Packet size: 512 bytes
Data rates: 10 Kbits

```
74    $ns_ node-config -adhocRouting blackholeAODV
75    set node_(5) [$ns_ node]
76    $ns_ at 0.0 "$node_(5) label \"BlackHole Node\""
77
78
79    $ns_ node-config -adhocRouting AODV
80    set node_(0) [$ns_ node]
81    set node_(1) [$ns_ node]
82    set node_(2) [$ns_ node]
83    set node_(3) [$ns_ node]
84    set node_(4) [$ns_ node]
85    set node_(6) [$ns_ node]
86    set node_(7) [$ns_ node]
87    set node_(8) [$ns_ node]
88    set node_(9) [$ns_ node]
89    set node_(10) [$ns_ node]
90    set node_(11) [$ns_ node]
91    set node_(12) [$ns_ node]
92    set node_(13) [$ns_ node]
93    set node_(14) [$ns_ node]
94    set node_(15) [$ns_ node]
95    set node_(16) [$ns_ node]
96    set node_(17) [$ns_ node]
97    set node_(18) [$ns_ node]
98    set node_(19) [$ns_ node]
```

*Figure 4.12 The statements for creating mobile nodes.*

And it did not use random packets in the simulation. Nodes in the simulation are generated by the statements that are shown in Fig. 4.12. The first statement: "$ns_ node-config -adhocRouting blackholeAODV" changes AODV routing protocol to "blackholeAODV" that is implemented in NS, so we can easily add the black hole AODV behavior with writing the number of node we wish to be black hole. And second statement: "$ns_ node-config -adhocRouting AODV" creates AODV nodes (Fig. 4.13).

Our simulation files are named with their simulation number and "*BlackHole*" number, for example, "sim1for1BlackHole6Mobility.tcl," "sim1for2BlackHole6Mobility.tcl". To run the AODV simulation without black hole attack, we changed the "*$val(nnaodv)*" variable to 6, 20, 30, 40, 50, 60, 70, 80, 90, and 100 nodes in different scenarios and then put the comment "#" in front of the "*$ns_ node-config -adhocRouting blackholeAODV*" statement and then we copied the Tcl script in same directory changing "*BlackHole*" definition of the file name with "*simple-AODV*," for example, "*simple-AODV_6nodees.tcl*"

```
# ===========================================================================
# Define options
# ===========================================================================
set val(chan) Channel/WirelessChannel ;        #ChannelType
set val(prop) Propagation/TwoRayGround ;       # radio-propagation model
set val(netif) Phy/WirelessPhy ;               # network interface type
set val(mac) Mac/802_11 ;                      # MAC type
set val(ifq) Queue/DropTail/PriQueue ;         # interface queue type
set val(ll) LL ;                               # link layer type
set val(ant) Antenna/OmniAntenna ;             # antenna model
set val(ifqlen) 150 ;                          # max packet in ifq
set val(nn) 20 ;                               # total number of mobilenodes
set val(n19) 20;                               # create 20 nodes
set val(nnaodv) 20 ;                           # number of AODV mobilenodes
set val(rp) AODV ;                             # routing protocol
```

*Figure 4.13 Mobile node configurations.*

is used for AODV without black hole attack including six nodes. The content of the first simulation file of the black hole AODV is shown in Appendix A. Node 5 being a black hole AODV node absorbs the packets in the connection from node 0 to node 4. Figs. 4.14–4.16 show how the black hole AODV node attracts the traffic. In Fig. 4.14, node 5 is black hole node, node 0 is sending node, and node 3 is destination node. Node 5 sends RREP to sending node's RREQ and after receiving data packets drop them:

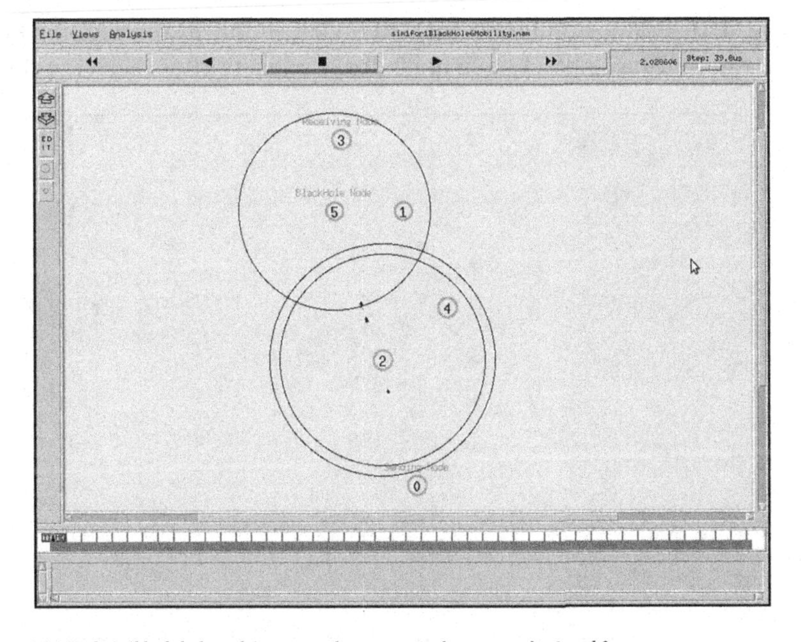


*Figure 4.14 Node 5 (black hole node) attracts the connection between nodes 0 and 3.*

Fig. 4.15 shows same simulation with 20 nodes; in this examination, it's supposed that node 19 is black hole node, node 0 is sending node, and node 4 is receiving node. As you can see, node 19 as a black hole node drops packets after receiving them:

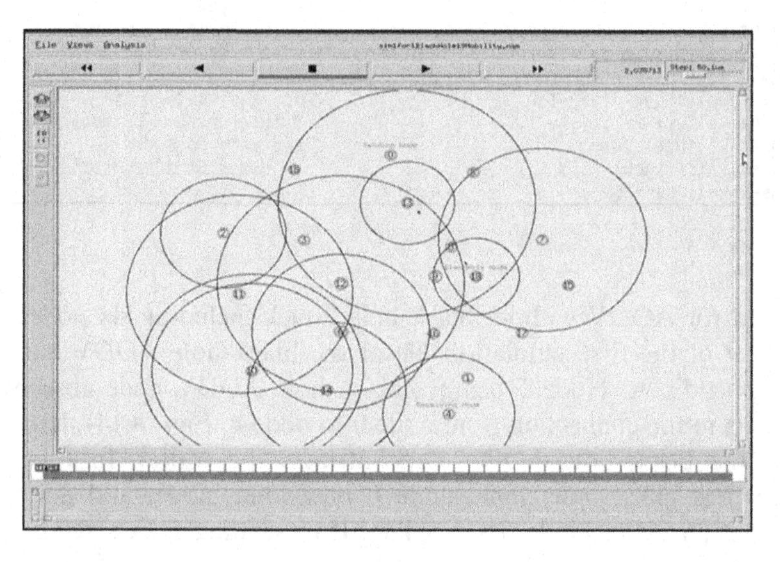

*Figure 4.15 Node 19 (black hole node) attracts the connection between nodes 0 and 4.*

Fig. 4.16 shows simulation with 30 nodes; in this examination as you can see node 29 as a black hole node drops packets after receiving them:

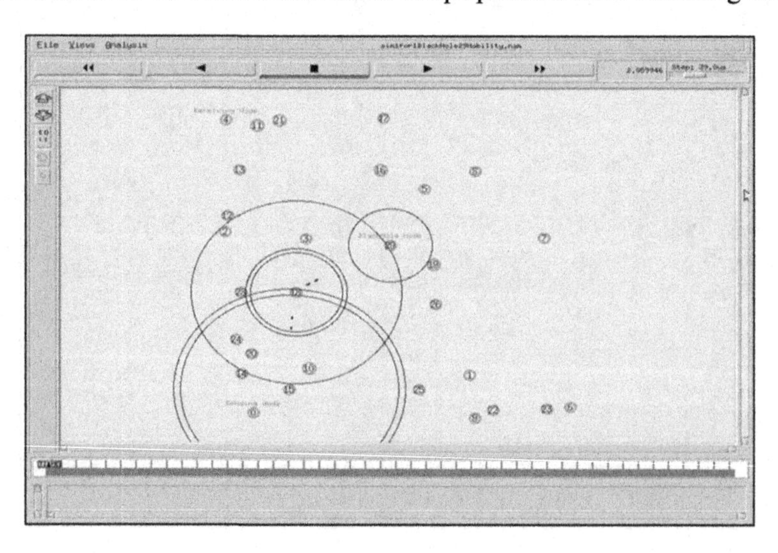

*Figure 4.16 Node 29 (black hole node) attracts the connection between nodes 0 and 4.*

### 4.4.2 Testing the Trace File and Obtaining the Results

It has gotten the simulation result from output and finds documents of the Tcl scripts, which has ".tr" elongation. Then, it finds documents that include all events in the simulation such as "when the packets are dispatched?," "which node generated them?," "which node has received?," "which type of package is dispatched?," "if it is dropped why it is dropped?," etc. In our simulations, we use "new-trace" document format that is particularly utilized in wireless networks and encompasses comprehensive happening information. The new-trace file experiment is shown in Appendix B. Its areas are explained in Appendix C.

To get the outcomes from the find files, we wanted only the event kind in area 0; node id (-Ni) and find the grade (-Nl) in area 4; and main address, destination address, and package sort in the area 5. To spot the on peak of facts and figures from the find document, we tend to use "cat" order of UNIX scheme and composed its outputs to a file for all find documents of the simulations. Of all the yields, we solely need "s" value of the happening facts and figures inside the area none, to count what number "cbr" packets "ar" dispatched by the causation node "r" worth of the happening data within the area 0, to enumerate how numerous CBR packets are obtained by the obtaining node "node id" worth of the node id information within the area 4, for the causation nodes or receiving nodes "MAC" worth of the find level information inside the area 4, to filter "MAC" level.

"Source Address" is values of the source and "Destination Address" is place visited by the address data in field 5, to count the packets that go from the dispatching node to the getting node. "cbr" value of the packet kind information within the field 5 to filter "cbr" packets.

To filter this information, we have a tendency to use "grep" command of the UNIX operating system reading the file developed by "cat" order and provided its output to word count (WC) order of the UNIX operating system as an input to count what quantity data have filtered and composed the end result to a new file. For demonstration; to count cbr packets the outcome by node 0 (sending node) the order "grep "s zero 0 --- 0.0 1.0 cbr" sim1forBlackHole.txt | wc -l >> result.txt" is utilized.

On other hand, to calculate "cbr" packets obtained by node 1 (receiving node), "*grep 's 0 MAC --- 0.0 1.0 cbr' sim1forBlackHole.txt* | *wc -l >> result.txt*" is used. These commands are directed for all nodes within the all replication and are in writing as a batch file. Content of this file is shown in Appendix C.

### 4.4.3 Appraisal of Results

Three different simulations have examined. In the first one, every node is working in cooperation with each other to keep the network in communication. The second one has one malicious node and third one has two malicious nodes that carry out by black hole attack. In this study, the results of these three simulations are compared to understand the network and node behaviors.

First it evaluated the packet loss. We counted how many packets are sent by the sending nodes and how many of them reached the receiving nodes. Tables 4.1–4.4 compare the normal and black hole network for total drop packets, end-to-end delay, routing overhead, and packet delivery ratio.

As we can see from Table 4.1, total drop packets of the black hole AODV is exaggerated more than the conventional AODV network simulations in all situations. We perceive from Table 4.1 that the packet loss already exists within the network. Also this is a result of packet drop at the node interface queue as a result of the density of knowledge traffic. Thus, we tend to alter node and packet parameters to reduce the information traffic. Desperate to evaluate the black hole impact within the network, we have to reduce the packet loss that happens at the network, except the black hole. In an exceedingly wireless ad hoc network that does not have any black hole, the information traffic could be dense and packets would possibly get lost, for example, in FTP traffic. Therefore, the information loss does not continuously say there was a black hole node within the network.

**Table 4.1 Total Drop Packets Comparison**

| Number of nodes | 6 | 20 | 30 | 40 | 50 | 60 | 70 | 80 | 90 | 100 |
|---|---|---|---|---|---|---|---|---|---|---|
| Before black hole attack | 0 | 0 | 0 | 0 | 0 | 0 | 0 | 0 | 0 | 0 |
| After one black hole attack | 498 | 498 | 498 | 498 | 498 | 498 | 498 | 498 | 498 | 498 |
| After two black hole attacks | 498 | 498 | 498 | 498 | 498 | 498 | 498 | 498 | 497 | 498 |

In Fig. 4.17, which is drawn from above table data, we can compare total drop packets before black hole attack and after one and two black hole attacks:

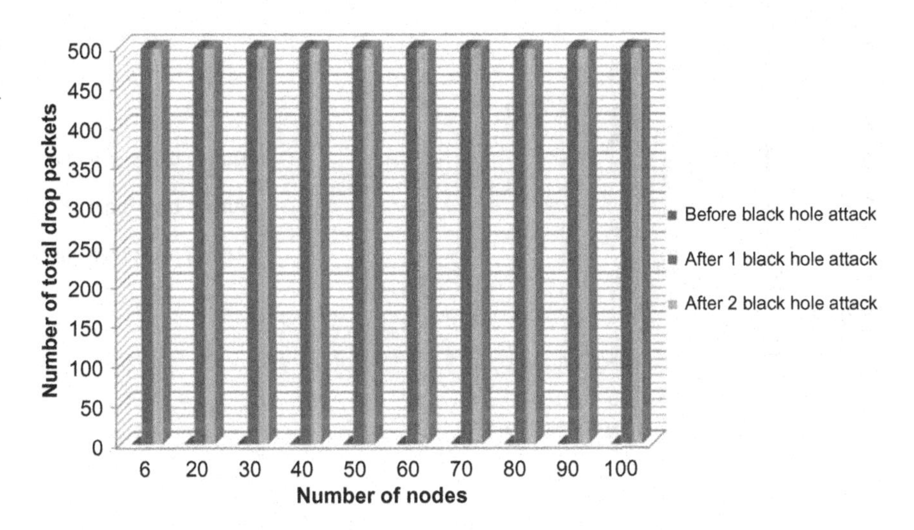


*Figure 4.17 Total drop packets comparison.*

Table 4.2 shows average end-to-end delay in different scenarios, as we can see average end-to-end delay increase in scenarios with black hole attack, it is because of dropping packets after receiving packets by black hole nodes.

**Table 4.2 End-to-End Delay Comparison**

| Number of nodes | 6 | 20 | 30 | 40 | 50 | 60 | 70 | 80 | 90 | 100 |
|---|---|---|---|---|---|---|---|---|---|---|
| Before black hole attack | 0.116364 | 0.116387 | 0.1059 | 0.101903 | 0.100342 | 0.100103 | 0.078637 | 0.078323 | 0.065856 | 0.0551 |
| After one black hole attack | 0 | 0.061414 | 0.0596 | 0 | 0 | 0 | 0.024322 | 0.0177599 | 0.017762 | 0 |
| After two black hole attacks | 0 | 0.023582 | 0.023923 | 0 | 0 | 0 | 0 | 0.017523 | 0.017482 | 0.017122 |

Fig. 4.18, which is drawn from above table's data, compares end-to-end delay before black hole attack and after one and two black hole attacks:

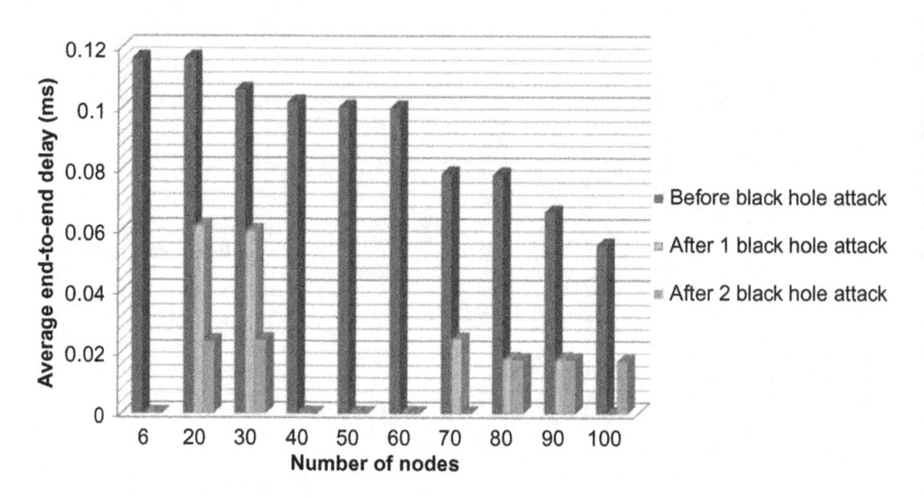


*Figure 4.18 End-to-end delay comparison.*

**Table 4.3 Routing Overhead Comparison**

| Number of nodes | 6 | 20 | 30 | 40 | 50 | 60 | 70 | 80 | 90 | 100 |
|---|---|---|---|---|---|---|---|---|---|---|
| Before black hole attack | 0.002 | 0.005 | 0.0089 | 0.01 | 0.015 | 0.014 | 0.016 | 0.024 | 0.026 | 0.029 |
| After one black hole attack | 0.001 | 0.01 | 0.009 | 0.016 | 0.017 | 0.018 | 0.02 | 0.024 | 0.026 | 0.03 |
| After two black hole attacks | 0.001 | 0.01 | 0.012 | 0.016 | 0.017 | 0.018 | 0.021 | 0.025 | 0.027 | 0.03 |

Table 4.3 shows routing request overhead for different scenarios. After entering black hole, nodes into the network routing request overhead increase; that means after black hole attack, number of data packets that transmit are less than number of control packets generate.

Fig. 4.19 is drawn from above table's data, which can compare routing overhead before black hole attack and after one and two black hole attacks:

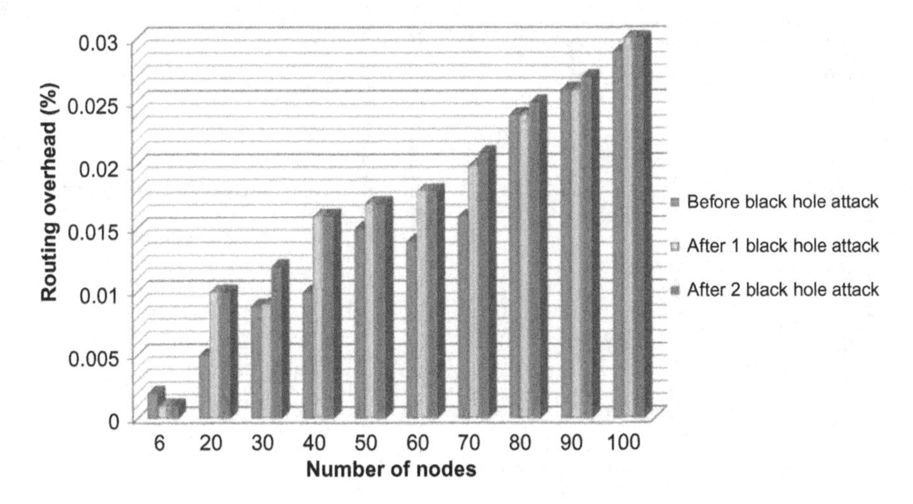


*Figure 4.19 Routing overhead comparison.*

Table 4.4 compares packet delivery ratio (PDR). As it shows, PDR is almost 1 before black hole attack that it means almost total packets sent by sender node are received by receiver node, but for network with black hole node PDR reduces to 0, that means almost whole of the packets sent by sender node are dropped by black hole nodes.

**Table 4.4 Packet Delivery Ratio Comparison**

| Number of nodes | 6 | 20 | 30 | 40 | 50 | 60 | 70 | 80 | 90 | 100 |
|---|---|---|---|---|---|---|---|---|---|---|
| Before black hole attack | 1 | 1 | 1 | 1 | 1 | 1 | 1 | 1 | 1 | 1 |
| After one black hole attack | 0 | 0 | 0 | 0 | 0 | 0 | 0 | 0 | 0 | 0 |
| After two black hole attacks | 0 | 0 | 0 | 0 | 0 | 0 | 0 | 0 | 0 | 0 |

Fig. 4.20 is drawn from above table's data, which can compare PDR before black hole attack and after one and two black hole attacks:

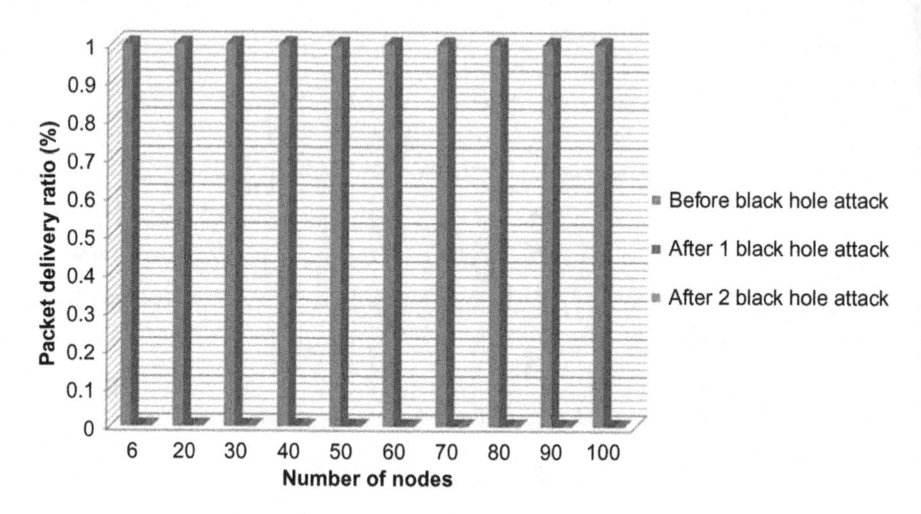


*Figure 4.20 Packet delivery ratio comparison.*

## 4.5 SUMMARY

This chapter investigated the performance of network before black hole attack, after single black hole attack, and after multiple black hole attacks. The output of this chapter is direct input for the next chapter. The purpose of this chapter is displaying network performance with single black hole attack and multiple black hole attacks and examining network performance under this attack.

CHAPTER 5

# Solution Execution and Result

## 5.1 INTRODUCTION

In previous chapter, it has implemented the black hole attack on AODV routing protocol and observed the effects of single and multiple attack on this protocol. In this chapter, we will go through to implementing IDSAODV as a solution versus this attack and will see its effects on AODV performance faces with single and multiple black hole attack.

## 5.2 AN OUTLINE OF INVESTIGATION

In previous chapter, it explained how black hole attack is implemented in NS2 and which results are obtained from the simulations. When we examine the trace file of the simulations that include one black hole node, we observe that after a while, second RREP message comes to source node from the real intermediate node.

To figure out how the second packet came to source node, it has created a simulation scenario with node positions as shown in Fig. 5.1. In the scenario, node 0 is the sending node, node 1 is black hole node, and node 4 is the receiving node. As the black hole send an RREP message without checking the tables, we assume that it is more likely for the first RREP to arrive from the black hole. In some cases, this idea may not work. For instance; the second RREP can be received at source node from an intermediate node which has fresh enough information about the destination node or the second RREP message may also come from the black hole node if the real destination node is nearer than the black hole node or in networks with multiple black hole nodes, second RREP can receive from other black hole nodes. These examples are extendable according to node condition in the network topology. In this book, it is tried to find out how IDSAODV eliminates the black hole effects in AODV with single black hole node and AODV with multiple black hole nodes, and if it improves the network performance or not.

```
280    void
281   idsAODV::rrep_insert(nsaddr_t id) {
282        idsBroadcastRREP *r = new idsBroadcastRREP(id);
283
284        assert(r);
285        r->expire = CURRENT_TIME + BCAST_ID_SAVE;
286        r->count ++;
287        LIST_INSERT_HEAD(&rrephead, r, link);
288   }
289
290    /* SRD */
291    idsBroadcastRREP *
292   idsAODV::rrep_lookup(nsaddr_t id) {
293        idsBroadcastRREP *r = rrephead.lh_first;
294
295        for( ; r; r = r->link.le_next) {
296            if (r->dst == id)
297                return r;
298        }
299        return NULL;
300   }
301
302
303    void
304   idsAODV::rrep_remove(nsaddr_t id) {
305        idsBroadcastRREP *r = rrephead.lh_first;
306
307        for( ; r; r = r->link.le_next) {
308            if (r->dst == id)
309                LIST_REMOVE(r,link);
310            delete r;
311            break;
312        }
313   }
314
315    void
316   idsAODV::rrep_purge() {
317        idsBroadcastRREP *r = rrephead.lh_first;
318        idsBroadcastRREP *rn;
319        double now = CURRENT_TIME;
320
321        for(; r; r = rn) {
322            rn = r->link.le_next;
323            if(r->expire <= now) {
324                LIST_REMOVE(r,link);
325                delete r;
326            }
327        }
328   }
```

*Figure 5.1 RREP caching mechanism.*

## 5.3 EXECUTION OF THE SOLUTION IN NS-2

To evaluate effects of the proposed solution, first, it needs to be implemented in NS-2. Therefore, should simulate the "AODV" protocol, changing it to "IDSAODV" as it did "blackholeAODV" before. To implement, the black hole has been changed the receive RREP function (recvRequest) of the blackholeaodv.cc file but to implement the solution had to change the receive RREP function (recvReply) and create RREP caching mechanism to count the second RREP message.

Fig. 5.1 shows the RREP caching mechanism. The "rrep_insert" function is for adding RREP messages, "rrep_lookup" function is for looking any RREP message up if it exists, "rrep_remove" function is for removing any record for RREP message that arrived from defined node and "rrep_purge" function is to delete periodically from the list if it has expired. It has chosen this expire time "*BCAST_ID_SAVE*" as 6 (mean seconds).

In the "recvReply" function, first we control if the RREP message arrived for itself and if it did, function looks the RREP message up if it has already arrived. If it did not, it inserts the RREP message for its destination address and returns from the function. If the RREP message is cached before for the same destination address, normal RREP function is carried out. Afterwards, if the RREP message is not meant for itself the node forwards the message to its appropriate neighbor. Fig. 5.2 shows how the receive RREP message function of the IDSAODV is carried out.

```
896
897     void
898   idsAODV::recvReply(Packet *p) {
899     //struct hdr_cmn *ch = HDR_CMN(p);
900         struct hdr_ip *ih = HDR_IP(p);
901         struct hdr_aodv_reply *rp = HDR_AODV_REPLY(p);
902         idsaodv_rt_entry *rt;
903         char suppress_reply = 0;
904         double delay = 0.0;
905         int count;
906
907         idsBroadcastRREP *r = rrep_lookup(rp->rp_dst);
908
909   #ifdef DEBUG
910         fprintf(stderr, "%d - %s: received a REPLY\n", index, __FUNCTION__);
911   #endif // DEBUG
912
913   #if 0
914         if (ih->daddr() == index) {
915             if (r == NULL) {
916                 rrep_insert(rp->rp_dst);
917                 Packet::free(p);
918                 return;
919             } else
920                 rrep_remove(rp->rp_dst);
921         }
922   #endif
923
924         if (r == NULL) {
925             count = 0;
926             rrep_insert(rp->rp_dst);
927         } else {
928             r->count++;
929             count = r->count;
930         }
```

*Figure 5.2 Receive RREP function of the IDSAODV.*

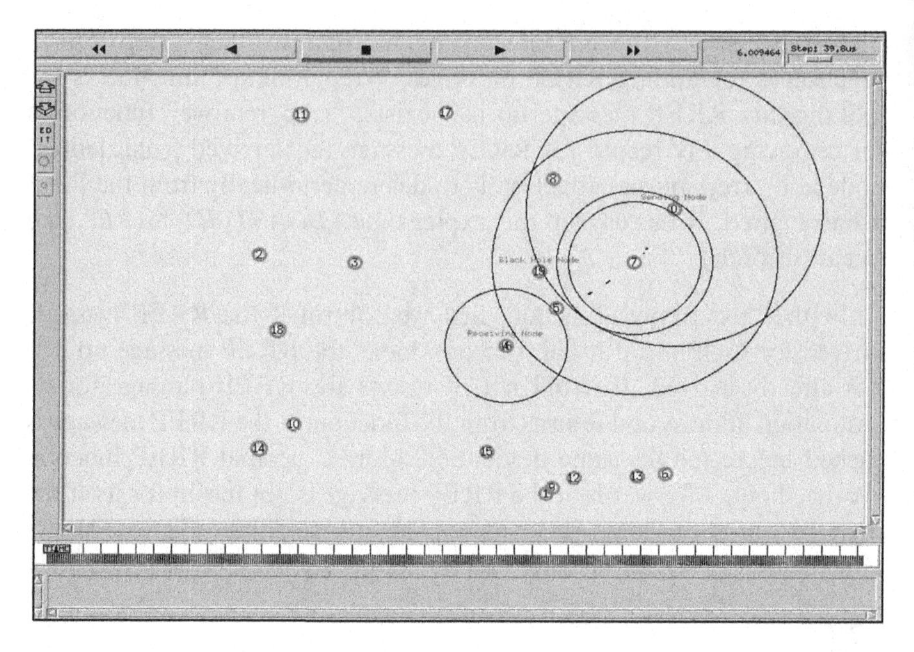

*Figure 5.3 CBR packet are reached to destination node properly.*

## 5.4 TESTING THE IDSAODV

For implementing the IDSAODV protocol in NS-2, we tried it in a Tcl simulation. In the scenario of the simulation, there are seven motionless nodes and node positions are the same as in the test simulation of the two RREP messages. In this simulation, IDSAODV protocol is used instead of AODV for all nodes except the black hole node (Node 1). To change the AODV protocol to IDSAODV, we only change "*$ns node-config-adhocRouting idsAODV*". When the simulation is compiled, we saw that sending node is sending the messages to receiving node properly. Fig. 5.3 shows that CBR packets are reaching the destination node as expected.

In the test simulation, it should be ensured that the IDSAODV implementation is correctly working. Then, the same simulations will be performed on the scenarios which have been used in Chapter 4 to compare the performance of IDS approach.

## 5.5 SIMULATION OF IDSAODV AND APPRAISAL OF RESULTS

To be able to evaluate if the proposed solution has been successful, we used same scenarios and simulation parameters as described in

Chapter 4 and also to be able to obtain the simulation results, it has been used a similar batch file adapted for idsaodv. Tables 5.1–5.8 compare effects of IDSAODV network with single and multiple black hole networks. Total drop packet for single black hole as we can see in Table 5.1 is almost zero.

**Table 5.1 Total Drop Packets Comparison**

| Number of nodes | 6 | 20 | 30 | 40 | 50 | 60 | 70 | 80 | 90 | 100 |
|---|---|---|---|---|---|---|---|---|---|---|
| Before black hole attack | 0 | 0 | 0 | 0 | 0 | 0 | 0 | 0 | 0 | 0 |
| After one black hole attack | 498 | 498 | 498 | 498 | 498 | 498 | 498 | 498 | 498 | 498 |
| After IDSAODV solution | 1 | 1 | 0 | 0 | 0 | 0 | 0 | 0 | 0 | 0 |

Fig. 5.4 is drawn from above table's data which can compare total drop packets before black hole attack, after one black hole attack and after using IDSAODV solution:

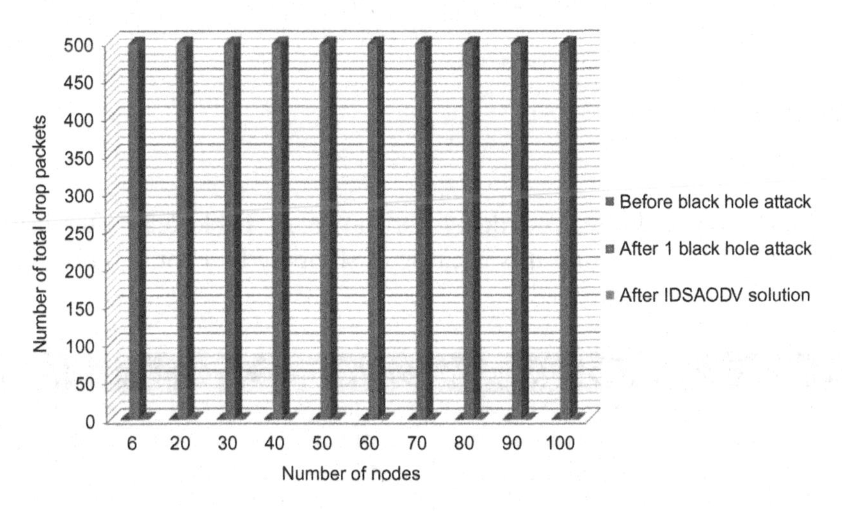


*Figure 5.4 Total drop packets comparison.*

As shown in Table 5.2, total drop packet for multiple black hole attacks after using IDSAODV is almost zero too.

**Table 5.2 Total Drop Packets Comparison**

| Number of nodes | 6 | 20 | 30 | 40 | 50 | 60 | 70 | 80 | 90 | 100 |
|---|---|---|---|---|---|---|---|---|---|---|
| Before black hole attack | 0 | 0 | 0 | 0 | 0 | 0 | 0 | 0 | 0 | 0 |
| After two black hole attack | 498 | 498 | 498 | 498 | 498 | 498 | 498 | 498 | 498 | 498 |
| After IDSAODV solution | 0 | 1 | 1 | 1 | 1 | 1 | 1 | 1 | 1 | 1 |

Fig. 5.5 is drawn from above table's data which can compare total drop packets before black hole attack, after two black hole attacks and after using IDSAODV solution:

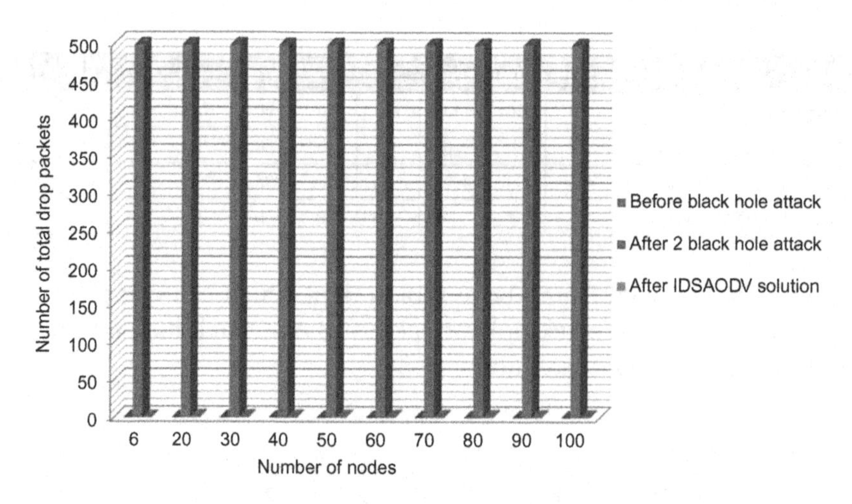


*Figure 5.5 Total drop packets comparison.*

But average end-to-end delay (as shown in Table 5.3) after using IDSAODV increase and it is because of skipping first route reply.

**Table 5.3 End-to-End Delay Comparison**

| Number of nodes | 6 | 20 | 30 | 40 | 50 | 60 | 70 | 80 | 90 | 100 |
|---|---|---|---|---|---|---|---|---|---|---|
| Before black hole attack | 0.116364 | 0.11639 | 0.1059 | 0.101903 | 0.100342 | 0.100103 | 0.078637 | 0.078323 | 0.065856 | 0.0551 |
| After one black hole attack | 0 | 0.06141 | 0.0596 | 0 | 0 | 0 | 0.024322 | 0.01776 | 0.017762 | |
| After IDSAODV solution | 0.216364 | 0.21639 | 0.2059 | 0.201903 | 0.200342 | 0.200103 | 0.178637 | 0.178323 | 0.165856 | 0.1551 |

Fig. 5.6 is drawn from above table's data which can compare end-to-end delay before black hole attack, after one black hole attack and after using IDSAODV solution:

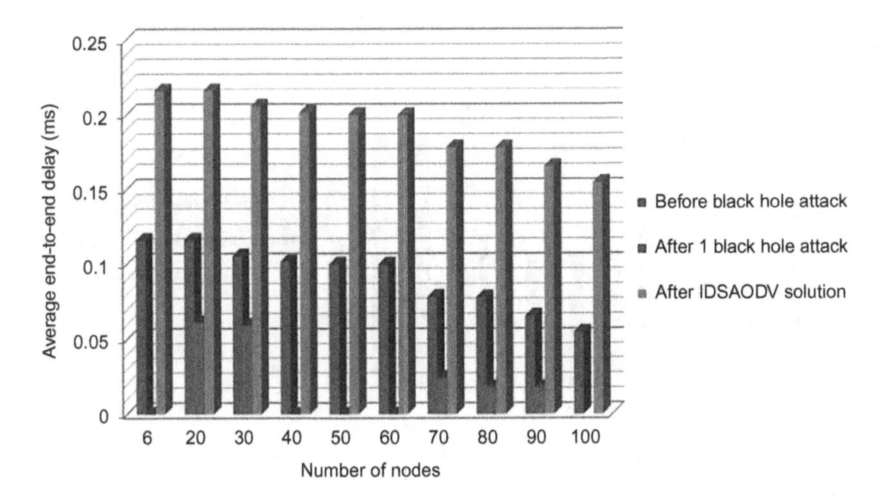


*Figure 5.6 End-to-end delay comparison.*

Same problem we have for end-to-end delay with IDSAODV for multiple black hole attacks, Table 5.4 shows the data came out from simulation:

**Table 5.4 End-to-End Delay Comparison**

| Number of nodes | 6 | 20 | 30 | 40 | 50 | 60 | 70 | 80 | 90 | 100 |
|---|---|---|---|---|---|---|---|---|---|---|
| Before black hole attack | 0.116364 | 0.116387 | 0.1059 | 0.101903 | 0.100342 | 0.100103 | 0.078637 | 0.078323 | 0.065856 | 0.0551 |
| After two black hole attack | 0 | 0.023582 | 0.02392 | 0 | 0 | 0 | 0 | 0.17523 | 0.017482 | 0.018422 |
| After IDSAODV solution | 0.226364 | 0.226387 | 0.2159 | 0.211903 | 0.210342 | 0.210103 | 0.188637 | 0.188323 | 0.175856 | 0.1651 |

Fig. 5.7 is drawn from above table's data which can compare end-to-end delay before black hole attack, after two black hole attacks and after using IDSAODV solution.

Routing request overhead after using IDSAODV decrease a little bit compared with black hole AODV but it is higher than AODV before black hole attack and it is because of increasing routing request compared with AODV because of skipping first route reply. We can observe the data came out from routing overhead simulation in Table 5.5.

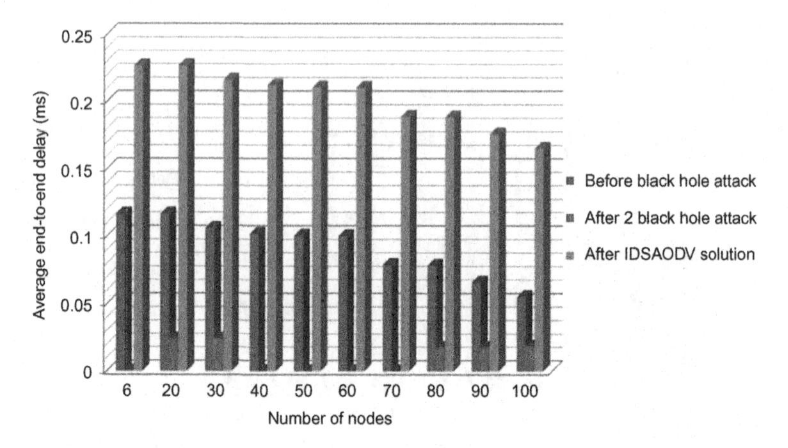


*Figure 5.7 End-to-end delay comparison.*

**Table 5.5 Routing Overhead Comparison**

| Number of nodes | 6 | 20 | 30 | 40 | 50 | 60 | 70 | 80 | 90 | 100 |
|---|---|---|---|---|---|---|---|---|---|---|
| Before black hole attack | 0.002 | 0.005 | 0.0079 | 0.01 | 0.015 | 0.014 | 0.016 | 0.024 | 0.026 | 0.029 |
| After two black hole attack | 0.001 | 0.01 | 0.009 | 0.016 | 0.017 | 0.018 | 0.02 | 0.026 | 0.028 | 0.03 |
| After IDSAODV solution | 0.003 | 0.007 | 0.008 | 0.012 | 0.015 | 0.016 | 0.019 | 0.024 | 0.026 | 0.029 |

Fig. 5.8 is drawn from above table's data which can compare routing overhead before black hole attack, after one black hole attack and after using IDSAODV solution:

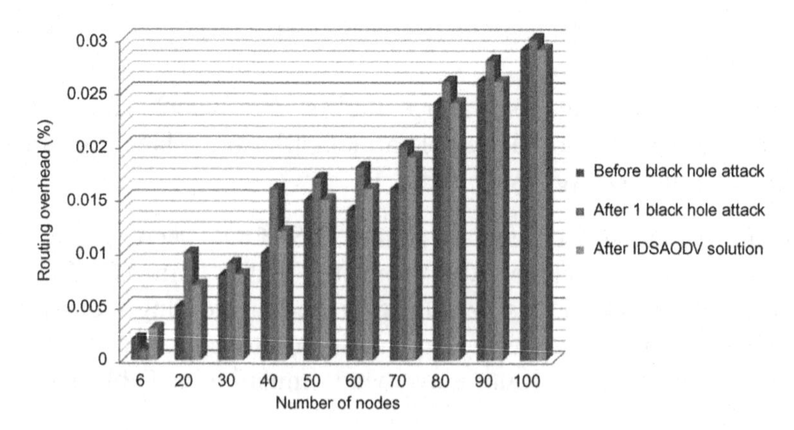


*Figure 5.8 Routing overhead comparison.*

Table 5.6 shows the result of routing overhead simulation after using IDSAODV solution for multiple black hole attacks:

**Table 5.6 Routing Overhead Comparison**

| Number of nodes | 6 | 20 | 30 | 40 | 50 | 60 | 70 | 80 | 90 | 100 |
|---|---|---|---|---|---|---|---|---|---|---|
| Before black hole attack | 0.002 | 0.005 | 0.0089 | 0.01 | 0.015 | 0.014 | 0.016 | 0.024 | 0.026 | 0.029 |
| After two black hole attack | 0.004 | 0.01 | 0.012 | 0.016 | 0.017 | 0.018 | 0.021 | 0.025 | 0.027 | 0.03 |
| After IDSAODV solution | 0.003 | 0.007 | 0.01 | 0.01 | 0.016 | 0.016 | 0.018 | 0.024 | 0.026 | 0.029 |

Fig. 5.9 is drawn from above table's data which can compare routing overhead before black hole attack, after two black hole attacks and after using IDSAODV solution:

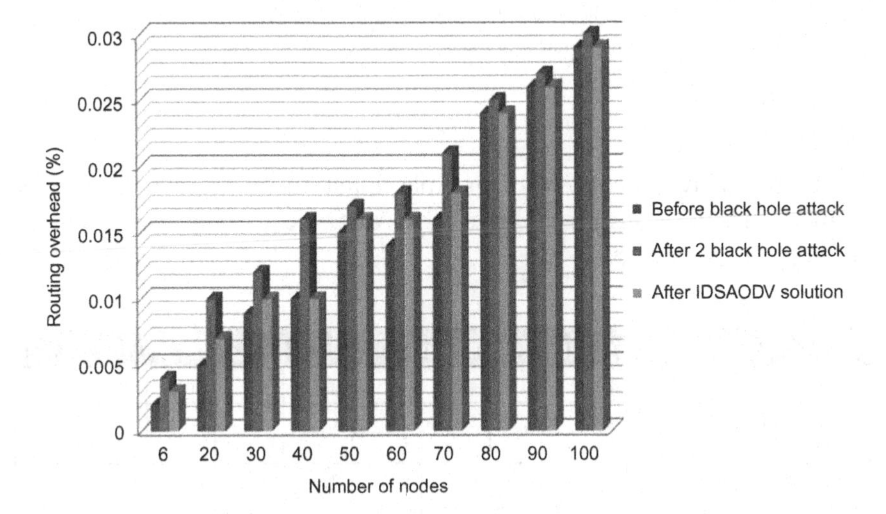


*Figure 5.9 Routing overhead comparison.*

In Table 5.7, you can see which packet delivery ratio after IDSAODV is equaled one, and it means almost all of the packets that have been sent by sender, received by destination node:

**Table 5.7 Packet Delivery Ratio Comparison**

| Number of nodes | 6 | 20 | 30 | 40 | 50 | 60 | 70 | 80 | 90 | 100 |
|---|---|---|---|---|---|---|---|---|---|---|
| Before black hole attack | 1 | 1 | 1 | 1 | 1 | 1 | 1 | 1 | 1 | 1 |
| After two black hole attack | 0 | 0 | 0 | 0 | 0 | 0 | 0 | 0 | 0 | 0 |
| After IDSAODV solution | 1 | 0.998 | 1 | 1 | 1 | 1 | 1 | 1 | 1 | 1 |

Fig. 5.10 is drawn from above table's data which can compare packet delivery ratio before black hole attack, after one black hole attack and after using IDSAODV solution:

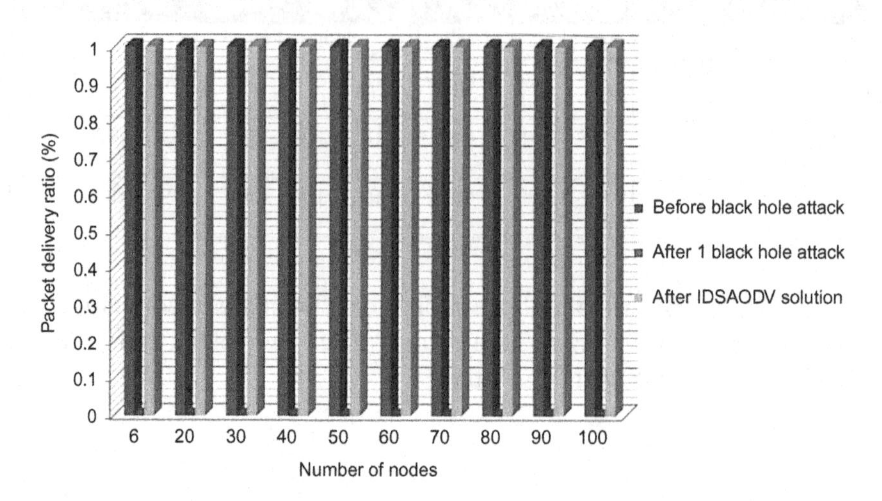


*Figure 5.10 Packet delivery ratio comparison.*

Packet delivery ratio for multiple black hole attacks also after using IDSAODV improved to almost one, we can see the result of simulation in Table 5.8:

**Table 5.8 Packet Delivery Ratio Comparison**

| Number of nodes | 6 | 20 | 30 | 40 | 50 | 60 | 70 | 80 | 90 | 100 |
|---|---|---|---|---|---|---|---|---|---|---|
| Before black hole attack | 1 | 1 | 1 | 1 | 1 | 1 | 1 | 1 | 1 | 1 |
| After two black hole attack | 0 | 0 | 0 | 0 | 0 | 0 | 0 | 0 | 0 | 0 |
| After IDSAODV solution | 1 | 0.998 | 0.998 | 0.998 | 0.998 | 0.998 | 0.998 | 0.998 | 0.998 | 0.998 |

Fig. 5.11 is drawn from above table's data which can compare packet delivery ratio before black hole attack, after two black hole attacks and after using IDSAODV solution:

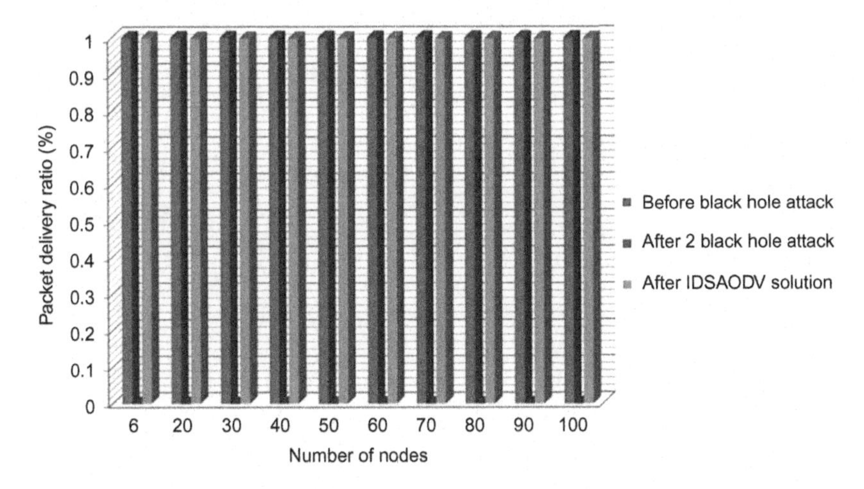


*Figure 5.11 Packet delivery ratio comparison.*

## 5.6 SUMMARY

In this chapter, we simulated IDSAODV solution for observing its effects on multiple black hole AODV performance, after that compared the results of this simulation with multiple black hole AODV. According the results of this chapter, IDSAODV can improve performance of black hole AODV.

# CHAPTER 6

# Conclusion and Future Work

## 6.1 OVERVIEW

In this study, we analyzed effect of the black hole in an AODV network. For this purpose, we implemented an AODV protocol that behaves as black hole in NS-2.35. We simulated 9 scenarios in this book on 6, 20, and 30 nodes, where the simulation was done with UDP and not TCP packets. In every scenario, we implemented network performance after and before multiple black hole attacks. After that, simulated IDSAODV solution with same scenarios to observe AODV behavior versus multiple black hole attacks under this proposed solution. First, investigated effects of black hole attack on network performance, which this attack increase number of drop packets and decrease packet delivery ratio, even with adding the number of black hole nodes, drop packets increase more, and packet delivery ratio drops off. After implementing IDSAODV on the network, drop packets decreased and packet delivery ratio improved. Other benefits of this solution are that the proposed solution requests minimum modification on AODV. It does not change packet format and can work together with AODV protocol and does not request additional routing overhead.

Black hole attack increases number of drop packets and decrease packet delivery ratio on MANET performance. After applying multiple numbers of black hole nodes on the network, drop packets will be more increased and packet delivery ratio drops off. Implementing IDSAODV on the network decreases drop packets and improves packet delivery ratio. IDSAODV solution requests minimum modification on AODV. IDSAODV does not change packet format and can work together with AODV protocol and does not request additional routing overhead.

## 6.2 CONTRIBUTION

The contributions of this book are listed as below:

- Effects of single black hole attack and multiple black hole attacks on MANET performance have investigated for networks with 10 different numbers of nodes.
- IDSAODV as a solution for black hole attack on MANET with single black hole attack and multiple black hole attacks for networks with 10 different numbers of nodes have implemented.
- Impacts of IDSAODV on MANET performance with multiple black hole attacks compared with its impacts on single black hole attack.

## 6.3 FUTURE WORK

According to our research, IDSAODV improve PDR and total drop packet but it cannot improve enough end-to-end delay and routing overhead against multiple black hole attacks, so future work will expand our research on other solutions that can more improve these parameters versus multiple black hole attacks.

# REFERENCES


Abolhasan, M., Wysocki, T., & Dutkiewicz, E. (2004). A review of routing protocols for mobile ad hoc networks. *Ad Hoc Networks*, *2*(1), 1–22.

Ahmad, Z., Manan, J.A., & Jalil, K.A. (2012). Performance evaluation on modified AODV protocols. Paper presented at: The Applied Electromagnetics (APACE), 2012 IEEE Asia-Pacific Conference.

Arunmozhi, S. A., & Venkataramani, Y. (2012). Black hole attack detection and performance improvement in mobile ad-hoc network. *Information Security Journal: A Global Perspective*, *21* (3), 150–158.

Arya, M., & Jain, Y. K. (2011). Grayhole attack and prevention in mobile ad hoc network. *International Journal of Computer Applications*, *27*(10), 21–26.

Bakshi, A., Sharma, A.K., & Mishra, A. (2013). Significance of mobile ad hoc networks (MANETS). *International Journal of Innovative Technology and Exploring Engineering (IJITEE)*, *2*(4).

Begam, U.S., & Murugaboopathi, G. (2013). A recent secure intrusion detection system for Manets. *International Journal of Emerging Technology and Advanced Engineering*, *3*, 45–62.

Bessire, M.L. (2006). System and method for ensuring the availability of a storage system: Google Patents.

Bhardwaj, P. K., Sharma, S., & Dubey, V. (2012). Comparative analysis of reactive and proactive protocol of mobile ad-hoc network. *International Journal*, *4*, 1281–1288.

Bhattacharyya, A., Banerjee, A., Bose, D., Saha, H. N., & Bhattacharya, D. (2011). Different types of attacks in mobile ad hoc network. *arXiv preprint arXiv*, 1111, 4090.

Bhushan, B., Gupta, S., & Nagpal, C. K. (2013). Comparison of on demand routing protocols. *International Journal of Information Technology*, *5*, 61–88.

Biswas, K., & Ali, M. L. (2007). Security threats in mobile ad hoc network. *Department of Interaction and System Design School of Engineering.*

Chavda, K.S., & Nimavat, A.V. (2013). Comparative analysis of detection and prevention techniques of black hole attack in AODV routing protocol of manet. *International Journal of Futuristic Science Engineering and Technology*, *1*(1).

Cheng, H. (2012). Genetic algorithms with hyper-mutation for dynamic load balanced clustering problem in mobile ad hoc networks. Paper presented at: The Natural Computation (ICNC), 2012 Eighth International Conference.

Chun, J., Shioura, A., Tien, T. M., & Tokuyama, T. (2013). *A unified view to greedy geometric routing algorithms in ad hoc networks algorithms for sensor systems* (pp. 54–65). Springer.

Dadhania, P., & Patel, S. (2013). Performance evaluation of routing protocol like AODV and DSR under black hole attacks. *Performance Evaluation*, *3*(1), 1487–1491.

Dangore, M.Y., & Sambare, S.S. (2013). A survey on detection of blackhole attack using AODV protocol in MANET. *International Journal on Recent and Innovation Trends in Computing and Communication*, *1*(1), 55–61.

Das, S.R., Belding-Royer, E.M., & Perkins, C.E. (2003). Ad hoc on-demand distance vector (AODV) routing. University of California, Santa Barbara.

Deng, H., Li, W., & Agrawal, D. P. (2002). Routing security in wireless ad hoc networks. *Communications Magazine, IEEE, 40*(10), 70–75.

Dokurer, S. (2006). *Simulation of black hole attack in wireless ad-hoc networks*. Ankara, Turkey: Atılım University.

Gupta, N., & Shrivastava, M. (2013). An evaluation of MANET routing protocol. *International Journal of Advanced Computer Research, 3*(1), 165–170.

Hassnawi, L. A., Ahmad, R. B., Yahya, A., Aljunid, S. A., & Elshaikh, M. (2012). Performance analysis of various routing protocols for motorway surveillance system cameras' network. *International Journal of Computer Science, 9*, 7–21.

Issariyakul, T. (2012). *Introduction to network simulator NS2*. New York: Springer Science + Business Media.

Issariyakul, T., & Hossain, E. (2012). *An introduction to network simulator-NS2*. New York: Springer.

Jawandhiya, P. M., Ghonge, M. M., Ali, M. S., & Deshpande, J. S. (2010). A survey of mobile ad hoc network attacks. *International Journal of Engineering Science and Technology, 2*(9), 4063–4071.

Kaur, D., & Kumar, N. (2013). Comparative analysis of AODV, OLSR, TORA, DSR and DSDV routing protocols in mobile ad-hoc networks. *International Journal, 5*, 39–46.

Kaur, R., & Rai, M. K. (2012). A novel review on routing protocols in MANETs. *Undergraduate Academic Research Journal (UARJ), 1*(1), 103–108.

Koyama, A., & Suzuki, H. (2013). Real object-oriented communication method for ad hoc networks. *Journal of Computer and System Sciences, 79*(7), 1101–1112.

Lee, S., Han, B., & Shin, M. (2002). Robust routing in wireless ad hoc networks. Paper presented at: The Parallel Processing Workshops, 2002. Proceedings. International Conference.

Liu, W., Nishiyama, H., Ansari, N., Yang, J., & Kato, N. (2013). Cluster-based certificate revocation with vindication capability for mobile ad hoc networks. *Parallel and Distributed Systems, IEEE Transactions on, 24*(2), 239–249.

Lu, S., Li, L., Lam, K.Y., & Jia, L. (2009). SAODV: A MANET routing protocol that can withstand black hole attack. Paper presented at: The Computational Intelligence and Security, 2009. CIS'09. International Conference.

Mahajan, V., Natu, M., & Sethi, A. (2008). Analysis of wormhole intrusion attacks in MANETS. Paper presented at: The Military Communications Conference, 2008. MILCOM 2008. IEEE.

Marti, S., Giuli, T.J., Lai, K., & Baker, M. (2000). Mitigating routing misbehavior in mobile ad hoc networks. Paper presented at: The Proceedings of the 6th Annual International Conference on Mobile Computing and Networking.

Menon, V. G., Johny, V., Tony, T., & Alias, E. (2013). Performance analysis of traditional topology based routing protocols in mobile ad hoc networks. *International Journal of Computer Science, 2*(01), 1–6.

Mistry, N., Jinwala, D.C., & Zaveri, M. (2010). Improving AODV protocol against blackhole attacks. Paper presented at: The Proceedings of the International Multiconference of Engineers and Computer Scientists.

Murthy, C. S. R., & Manoj, B. S. (2004). *Ad hoc wireless networks: Architectures and protocols*. Upper Saddle River, NJ: Prentice Hall.

Natarajan, K., & Mahadevan, G. (2013). A succinct comparative analysis and performance evaluation of MANET routing protocols. Paper presented at: The Computer Communication and Informatics (ICCCI), 2013 International Conference.

Nguyen, H. L., & Nguyen, U. T. (2008). A study of different types of attacks on multicast in mobile ad hoc networks. *Ad Hoc Networks*, *6*(1), 32–46.

Parsons, M., & Ebinger, P. (2009). Performance evaluation of the impact of attacks on mobile ad-hoc networks. Paper presented at: The Proceedings of Field Failure Data Analysis Workshop September 27–30, Niagara Falls, New York, USA.

Pegueno, G. A., & Rivera, J. R. (2006). *Extension to MAC 802.11 for performance improvement in MANET*. Sweden: Karlstads University.

Rafsanjani, M. K., Movaghar, A., & Koroupi, F. (2008). Investigating intrusion detection systems in MANET and comparing IDSs for detecting misbehaving nodes. *World Academy of Science, Engineering and Technology*, *44*, 351–355.

Raj, P. N., & Swadas, P. B. (2009). Dpraodv: A dyanamic learning system against blackhole attack in aodv based manet. *arXiv preprint arXiv*, *0909*, 2371.

Ramaswamy, S., Fu, H., Sreekantaradhya, M., Dixon, J., & Nygard, K.E. (2003). *Prevention of cooperative black hole attack in wireless ad hoc networks*. Paper presented at: The International Conference on Wireless Networks.

Refaei, M.T., Srivastava, V., DaSilva, L., & Eltoweissy, M. (2005). *A reputation-based mechanism for isolating selfish nodes in ad hoc networks*. Paper presented at: The Mobile and Ubiquitous Systems: Networking and Services, 2005. MobiQuitous 2005. The Second Annual International Conference.

Ros, F. J., & Ruiz, M. P. (2004). *Implementing a new manet unicast routing protocol in NS2*. Murcia, Spain.

Roy, D. B., Chaki, R., & Chaki, N. (2010). A new cluster-based wormhole intrusion detection algorithm for mobile ad-hoc networks. *arXiv preprint arXiv*, *1004*, 0587.

Ruzgar, E., & Dagdeviren, O. (2013). Performance evaluation of distributed synchronous greedy graph coloring algorithms on wireless ad hoc and sensor networks. *International Journal of Computer Networks & Communications (IJCNC)*, *5*(2), 169–179.

Saha, H.N., Bhattacharyya, D., Banerjee, P.K., Bhattacharyya, A., Banerjee, A., & Bose, D. (2012). Study of different attacks in MANET with its detection & mitigation schemes. *International Journal of Advanced Engineering Technology*, E-ISSN, 976–3945.

Shanthi, N., Lganesan, D. R., & Ramar, D. R. K. (2009). Study of different attacks on multicast mobile ad-hoc network. *Journal of Theoretical and Applied Information Technology*, *9*(2), 45–51.

Sharma, S., & Gupta, R. (2009). Simulation study of blackhole attack in the mobile Ad Hoc networks. *Journal of Engineering Science and Technology*, *4*(2), 243–250.

Singh, P.K., & Sharma, G. (2012). *An efficient prevention of black hole problem in AODV routing protocol* in MANET. Paper presented at: The Trust, Security and Privacy in Computing and Communications (TrustCom), 2012 IEEE 11th International Conference.

Singh, T. P., Dua, S., & Das, V. (2012). Energy-efficient routing protocols in mobile ad-hoc networks. *International Journal of Advanced Research in Computer Science and Software Engineering*, *5*(1).

Srivastava, M. (2012). A performance analysis of routing protocols in mobile ad-hoc networks. *International Journal of Engineering*, *1*(8).

Stajano, F., & Anderson, R. (2002). The resurrecting duckling: Security issues for ubiquitous computing. *Computer*, *35*(4), 22–26.

Student, V. R. P. G., & Dhir, R. (2013). A study of ad-hoc network: A review. *International Journal*, *3*(3), 135–138.

Sun, B., Guan, Y., Chen, J., & Pooch, U.W. (2003). *Detecting black-hole attack in mobile ad hoc networks*. Paper presented at: The Personal Mobile Communications Conference, 2003. 5th European (Conf. Publ. No. 492).

Tamilselvan, L., & Sankaranarayanan, V. (2006). *Solution to prevent rushing attack in wireless mobile ad hoc networks.* Paper presented at: The Ad Hoc and Ubiquitous Computing, 2006. ISAUHC'06. International Symposium.

Tsujii, S., & Itoh, T. (1989). An ID-based cryptosystem based on the discrete logarithm problem. *Selected Areas in Communications, IEEE Journal on*, *7*(4), 467–473.

Ullah, I., & Rehman, S.U. (2010). Analysis of black hole attack on MANETs using different MANET routing protocols. *A Mater Thesis, Electrical Engineering, Thesis No. MEE*, 10, 62.

Usha, U., & Bose, B. (2012). Comparing the impact of black hole and gray hole attacks in mobile adhoc networks. *Journal of Computer Science*, *8*(11), 2012.

Vinayakray-Jani, P. (2002). *Security within ad hoc networks.* Paper presented at: The First PAMPAS Workshop.

Vincent, S.S.M., & Meshach, W.T. (2012). Preventing black hole attack in MANETs using randomized multipath routing algorithm. *International Journal of Soft Computing and Engineering (IJSCE)*, *1*(ETIC2011), 30–33.

Wei, C., Xiang, L., Yuebin, B., & Xiaopeng, G. (2007). *A new solution for resisting gray hole attack in mobile ad-hoc networks.* Paper presented at: The Communications and Networking in China, 2007. CHINACOM'07. Second International Conference.

Zhu, C., Lee, M.J., & Saadawi, T. (2003). *Rtt-based optimal waiting time for best route selection in ad hoc routing protocols.* Paper presented at: The Military Communications Conference, 2003. MILCOM'03. 2003 IEEE.

# APPENDIX A

# Blackholeaodv TCL File

```
# blackholeaodv TCL file
#
# == == == == == == == == == == == == == == == == == == == =
# Define options
#
# == == == == == == == == == == == == == == == == == == == =
set val(Chan) Channel/WirelessChannel ;      #ChannelType
set val(prop) Propagation/TwoRayGround ;     # radio-propagation model
set val(netif) Phy/WirelessPhy ;             # network interface type
set val(mac) Mac/802_11 ;                    # MAC type
set val(ifq) Queue/DropTail/PriQueue ;       # interface queue type
set val(ll) LL ;                             # link layer type
set val(ant) Antenna/OmniAntenna ;           # antenna model
set val(ifqlen) 150 ;                        # max packet in ifq
set val(nn) 6 ;                          # total number of mobilenodes
set val(n6) 6;                           # create 6 nodes
set val(nnaodv) 5 ;                      # number of AODV mobilenodes
set val(rp) AODV ;                       # routing protocol
set val(energymodel)  EnergyModel ;      # Energy Model
set val(initialenergy)  100 ;            # value in joules
set val(x) 670 ;                         # X dimension of topography
set val(y) 670 ;                         # Y dimension of topography
set val(cstop) 500 ;                     # time of connections end
set val(stop) 500 ;                      # time of simulation end
# set val(cp) "scenarios/scen1forAODV-n20-t500-x750-y750" ;
#Connection pattern
# set val(cc) "scenarios/cbr" ;                #CBR Connections

# Initialize Global Variables
set ns_[new Simulator]
$ns_ use-newtrace
set tracefd[open sim1for1BlackHole6Mobility.tr w]
$ns_ trace-all $tracefd
set namtrace [open sim1for1BlackHole6Mobility.nam w]
$ns_ namtrace-all-wireless $namtrace $val(x) $val(y)

# set up topography object
set topo [new Topography]
$topo load_flatgrid $val(x) $val(y)
```

```
# Create God
create-god $val(nn)
# Create channel #1 and #2
set chan_1_[new $val(Chan)]
set chan_2_[new $val(Chan)]

# configure node, please note the change below.
$ns_ node-config -adhocRouting $val(rp) \
-llType $val(ll) \
-macType $val(mac) \
-ifqType $val(ifq) \
-ifqLen $val(ifqlen) \
-antType $val(ant) \
-propType $val(prop) \
-phyType $val(netif) \
-topoInstance $topo \
-energyModel $val(energymodel) \
-initialEnergy $val(initialenergy) \
-rxPower 35.28e-3 \
-txPower 31.32e-3 \
-idlePower 712e-6 \
-sleepPower 144e-9 \
-agentTrace ON \
-routerTrace ON \
-macTrace ON \
-movementTrace ON \
-channel $chan_1_ \

$ns_ node-config -adhocRouting AODV
set node_(0) [$ns_ node]
#set node_(1) [$ns_ node]
set node_(1) [$ns_ node]
set node_(2) [$ns_ node]
set node_(3) [$ns_ node]
set node_(4) [$ns_ node]

$ns_ node-config -adhocRouting blackholeAODV
#set node_(5) [$ns_ node]
#$ns_ at 0.0 "$node_(5) label \"BlackHole Node\""
#set node_(4) [$ns_ node]
#$ns_ at 0.0 "$node_(4) label \"BlackHole Node\""
set node_(5) [$ns_ node]
$ns_ at 0.0 "$node_(5) label \"BlackHole Node\""

# create 6 nodes
# define positions in X and Y axis
$node_(0) set X_ 345.0
$node_(0) set Y_ 201.0
$node_(0) set Z_ 0.0
$ns_ at 0.0 "$node_(0) label \"Sending Node\""
```

```
$node_(1) set X_ 330.0
$node_(1) set Y_ 500.0
$node_(1) set Z_ 0.0

$node_(2) set X_ 307.0
$node_(2) set Y_ 338.0
$node_(2) set Z_ 0.0

$node_(3) set X_ 261.0
$node_(3) set Y_ 578.0
$node_(3) set Z_ 0.0
$ns_ at 0.0 "$node_(3) label \"Receiving Node\""

$node_(4) set X_ 380.0
$node_(4) set Y_ 395.0
$node_(4) set Z_ 0.0

$node_(5) set X_ 253.0
$node_(5) set Y_ 500.0
$node_(5) set Z_ 0.0

# Now produce some simple node movements
$ns_ at 50.0 "$node_(1) setdest 25.0  20.0 10.0"
$ns_ at 10.0 "$node_(0) setdest 350.0 220.0 11.0"
$ns_ at 75.0 "$node_(2) setdest 125.0 270.0 12.0"
$ns_ at 37.0 "$node_(3) setdest 400.0  280.0 13.0"
$ns_ at 13.0 "$node_(4) setdest 261.0 560.0 14.0"
$ns_ at 43.0 "$node_(5) setdest 113.0  8.0 15.0"

$ns_ at 109.0 "$node_(1) setdest 254.0 450.0 10.0"
$ns_ at 300.0 "$node_(0) setdest 345.0 201.0 11.0"
$ns_ at 391.0 "$node_(2) setdest 490.0 181.0 12.0"
$ns_ at 320.0 "$node_(3) setdest 400.0 300.0 13.0"
$ns_ at 104.0 "$node_(4) setdest 261.0 578.0 14.0"
$ns_ at 200.0 "$node_(5) setdest  97.0 549.0 15.0"

# Setup traffic flow between nodes
#Set Udp, cbr agent and attach those with nodes

set udp1 [new Agent/UDP]
$ns_ attach-agent $node_(0) $udp1
$udp1 set class_ 0

#$udp1 set fid_ 2
set cbr1 [new Application/Traffic/CBR]
$cbr1 attach-agent $udp1
$cbr1 set packetSize_ 512
$cbr1 set rate_ 10kb
$cbr1 set interval_ 1

# Attach null agent for sink
set null1 [new Agent/Null]
```

```
$ns_ attach-agent $node_(3) $null1
$ns_ connect $udp1 $null1

# Start the traffic generator
$ns_ at 2.0 "$cbr1 start"

#Define initial node position in NAM for 6 nodes
$ns_ initial_node_pos $node_(0) 20
$ns_ initial_node_pos $node_(1) 20
$ns_ initial_node_pos $node_(2) 20
$ns_ initial_node_pos $node_(3) 20
$ns_ initial_node_pos $node_(4) 20
$ns_ initial_node_pos $node_(5) 20

# Tell all nodes when the simulation ends
for{ set i 0} { $i < $val(n6) } { incr i} {
$ns_ at $val(stop).000000001 "$node_($i) reset";
}

# Ending nam and simulation
$ns_ at $val(stop) "finish"
$ns_ at $val(stop).0 "$ns_ trace-annotate \"Simulation has ended\""
$ns_ at $val(stop).00000001 "puts \"NS EXITING...\"; $ns_ halt"
proc finish{} {
global ns_ tracefd namtrace
$ns_ flush-trace
close $tracefd
close $namtrace
exec nam sim1for1BlackHole6Mobility.nam &
exit 0
}

puts "Starting Simulation..."
$ns_ run
```

# APPENDIX B

# Trace File Example

```
s -t 2.000000000 -Hs 0 -Hd -2 -Ni 0 -Nx 345.00 -Ny 201.00 -Nz 0.00 -Ne
100.000000 -Nl AGT -Nw --- -Ma 0 -Md 0 -Ms 0 -Mt 0 -Is 0.0 -Id 3.0 -It cbr
-Il 512 -If 0 -Ii 0 -Iv 32 -Pn cbr -Pi 0 -Pf 0 -Po 0
r -t 2.000000000 -Hs 0 -Hd -2 -Ni 0 -Nx 345.00 -Ny 201.00 -Nz 0.00 -Ne
100.000000 -Nl RTR -Nw --- -Ma 0 -Md 0 -Ms 0 -Mt 0 -Is 0.0 -Id 3.0 -It cbr
-Il 512 -If 0 -Ii 0 -Iv 32 -Pn cbr -Pi 0 -Pf 0 -Po 0
s -t 2.000000000 -Hs 0 -Hd -2 -Ni 0 -Nx 345.00 -Ny 201.00 -Nz 0.00 -Ne
100.000000 -Nl RTR -Nw --- -Ma 0 -Md 0 -Ms 0 -Mt 0 -Is 0.255 -Id -1.255 -It
AODV -Il 48 -If 0 -Ii 0 -Iv 30 -P aodv -Pt 0x2 -Ph 1 -Pb 1 -Pd 3 -Pds 0 -Ps 0
-Pss 4 -Pc REQUEST
s -t 2.000535000 -Hs 0 -Hd -2 -Ni 0 -Nx 345.00 -Ny 201.00 -Nz 0.00 -Ne
100.000000 -Nl MAC -Nw --- -Ma 0 -Md ffffffff -Ms 0 -Mt 800 -Is 0.255 -Id
-1.255 -It AODV -Il 106 -If 0 -Ii 0 -Iv 30 -P aodv -Pt 0x2 -Ph 1 -Pb 1 -Pd 3
-Pds 0 -Ps 0 -Pss 4 -Pc REQUEST
N -t 2.000535 -n 2 -e 99.998546
N -t 2.000536 -n 4 -e 99.998546
N -t 2.000536 -n 1 -e 99.998546
N -t 2.000536 -n 5 -e 99.998546
N -t 2.000536 -n 3 -e 99.998546
r -t 2.001383474 -Hs 2 -Hd -2 -Ni 2 -Nx 307.00 -Ny 338.00 -Nz 0.00 -Ne
99.998546 -Nl MAC -Nw --- -Ma 0 -Md ffffffff -Ms 0 -Mt 800 -Is 0.255 -Id
-1.255 -It AODV -Il 48 -If 0 -Ii 0 -Iv 30 -P aodv -Pt 0x2 -Ph 1 -Pb 1 -Pd 3
-Pds 0 -Ps 0 -Pss 4 -Pc REQUEST
r -t 2.001383657 -Hs 4 -Hd -2 -Ni 4 -Nx 380.00 -Ny 395.00 -Nz 0.00 -Ne
99.998546 -Nl MAC -Nw --- -Ma 0 -Md ffffffff -Ms 0 -Mt 800 -Is 0.255 -Id
-1.255 -It AODV -Il 48 -If 0 -Ii 0 -Iv 30 -P aodv -Pt 0x2 -Ph 1 -Pb 1 -Pd 3
-Pds 0 -Ps 0 -Pss 4 -Pc REQUEST
r -t 2.001408474 -Hs 2 -Hd -2 -Ni 2 -Nx 307.00 -Ny 338.00 -Nz 0.00 -Ne
99.998546 -Nl RTR -Nw --- -Ma 0 -Md ffffffff -Ms 0 -Mt 800 -Is 0.255 -Id
-1.255 -It AODV -Il 48 -If 0 -Ii 0 -Iv 30 -P aodv -Pt 0x2 -Ph 1 -Pb 1 -Pd 3
-Pds 0 -Ps 0 -Pss 4 -Pc REQUEST
r -t 2.001408657 -Hs 4 -Hd -2 -Ni 4 -Nx 380.00 -Ny 395.00 -Nz 0.00 -Ne
99.998546 -Nl RTR -Nw --- -Ma 0 -Md ffffffff -Ms 0 -Mt 800 -Is 0.255 -Id
-1.255 -It AODV -Il 48 -If 0 -Ii 0 -Iv 30 -P aodv -Pt 0x2 -Ph 1 -Pb 1 -Pd 3
-Pds 0 -Ps 0 -Pss 4 -Pc REQUEST
s -t 2.001550521 -Hs 2 -Hd -2 -Ni 2 -Nx 307.00 -Ny 338.00 -Nz 0.00 -Ne
99.998546 -Nl RTR -Nw --- -Ma 0 -Md ffffffff -Ms 0 -Mt 800 -Is 2.255 -Id
```

```
-1.255 -It AODV -Il 48 -If 0 -Ii 0 -Iv 29 -P aodv -Pt 0x2 -Ph 2 -Pb 1 -Pd 3
-Pds 0 -Ps 0 -Pss 4 -Pc REQUEST
s -t 2.001945521 -Hs 2 -Hd -2 -Ni 2 -Nx 307.00 -Ny 338.00 -Nz 0.00 -Ne
99.998546 -Nl MAC -Nw --- -Ma 0 -Md fffffff -Ms 2 -Mt 800 -Is 2.255 -Id
-1.255 -It AODV -Il 106 -If 0 -Ii 0 -Iv 29 -P aodv -Pt 0x2 -Ph 2 -Pb 1 -Pd 3
-Pds 0 -Ps 0 -Pss 4 -Pc REQUEST
N -t 2.001946 -n 4 -e 99.998515
N -t 2.001946 -n 0 -e 99.998519
N -t 2.001946 -n 1 -e 99.998515
N -t 2.001946 -n 5 -e 99.998515
N -t 2.001946 -n 3 -e 99.998515
s -t 2.002289884 -Hs 4 -Hd -2 -Ni 4 -Nx 380.00 -Ny 395.00 -Nz 0.00 -Ne
99.998515 -Nl RTR -Nw --- -Ma 0 -Md fffffff -Ms 0 -Mt 800 -Is 4.255 -Id
-1.255 -It AODV -Il 48 -If 0 -Ii 0 -Iv 29 -P aodv -Pt 0x2 -Ph 2 -Pb 1 -Pd 3
-Pds 0 -Ps 0 -Pss 4 -Pc REQUEST
r -t 2.002793829 -Hs 4 -Hd -2 -Ni 4 -Nx 380.00 -Ny 395.00 -Nz 0.00 -Ne
99.998515 -Nl MAC -Nw --- -Ma 0 -Md fffffff -Ms 2 -Mt 800 -Is 2.255 -Id
-1.255 -It AODV -Il 48 -If 0 -Ii 0 -Iv 29 -P aodv -Pt 0x2 -Ph 2 -Pb 1 -Pd 3
-Pds 0 -Ps 0 -Pss 4 -Pc REQUEST
r -t 2.002793994 -Hs 0 -Hd -2 -Ni 0 -Nx 345.00 -Ny 201.00 -Nz 0.00 -Ne
99.998519 -Nl MAC -Nw --- -Ma 0 -Md fffffff -Ms 2 -Mt 800 -Is 2.255 -Id
-1.255 -It AODV -Il 48 -If 0 -Ii 0 -Iv 29 -P aodv -Pt 0x2 -Ph 2 -Pb 1 -Pd 3
-Pds 0 -Ps 0 -Pss 4 -Pc REQUEST
r -t 2.002794066 -Hs 1 -Hd -2 -Ni 1 -Nx 330.00 -Ny 500.00 -Nz 0.00 -Ne
99.998515 -Nl MAC -Nw --- -Ma 0 -Md fffffff -Ms 2 -Mt 800 -Is 2.255 -Id
-1.255 -It AODV -Il 48 -If 0 -Ii 0 -Iv 29 -P aodv -Pt 0x2 -Ph 2 -Pb 1 -Pd 3
-Pds 0 -Ps 0 -Pss 4 -Pc REQUEST
r -t 2.002794090 -Hs 5 -Hd -2 -Ni 5 -Nx 253.00 -Ny 500.00 -Nz 0.00 -Ne
99.998515 -Nl MAC -Nw --- -Ma 0 -Md fffffff -Ms 2 -Mt 800 -Is 2.255 -Id
-1.255 -It AODV -Il 48 -If 0 -Ii 0 -Iv 29 -P aodv -Pt 0x2 -Ph 2 -Pb 1 -Pd 3
-Pds 0 -Ps 0 -Pss 4 -Pc REQUEST
r -t 2.002794335 -Hs 3 -Hd -2 -Ni 3 -Nx 261.00 -Ny 578.00 -Nz 0.00 -Ne
99.998515 -Nl MAC -Nw --- -Ma 0 -Md fffffff -Ms 2 -Mt 800 -Is 2.255 -Id
-1.255 -It AODV -Il 48 -If 0 -Ii 0 -Iv 29 -P aodv -Pt 0x2 -Ph 2 -Pb 1 -Pd 3
-Pds 0 -Ps 0 -Pss 4 -Pc REQUEST
r -t 2.002818829 -Hs 4 -Hd -2 -Ni 4 -Nx 380.00 -Ny 395.00 -Nz 0.00 -Ne
99.998515 -Nl RTR -Nw --- -Ma 0 -Md fffffff -Ms 2 -Mt 800 -Is 2.255 -Id
-1.255 -It AODV -Il 48 -If 0 -Ii 0 -Iv 29 -P aodv -Pt 0x2 -Ph 2 -Pb 1 -Pd 3
-Pds 0 -Ps 0 -Pss 4 -Pc REQUEST
r -t 2.002818994 -Hs 0 -Hd -2 -Ni 0 -Nx 345.00 -Ny 201.00 -Nz 0.00 -Ne
99.998519 -Nl RTR -Nw --- -Ma 0 -Md fffffff -Ms 2 -Mt 800 -Is 2.255 -Id
-1.255 -It AODV -Il 48 -If 0 -Ii 0 -Iv 29 -P aodv -Pt 0x2 -Ph 2 -Pb 1 -Pd 3
-Pds 0 -Ps 0 -Pss 4 -Pc REQUEST
r -t 2.002819066 -Hs 1 -Hd -2 -Ni 1 -Nx 330.00 -Ny 500.00 -Nz 0.00 -Ne
99.998515 -Nl RTR -Nw --- -Ma 0 -Md fffffff -Ms 2 -Mt 800 -Is 2.255 -Id
-1.255 -It AODV -Il 48 -If 0 -Ii 0 -Iv 29 -P aodv -Pt 0x2 -Ph 2 -Pb 1 -Pd 3
-Pds 0 -Ps 0 -Pss 4 -Pc REQUEST
r -t 2.002819090 -Hs 5 -Hd -2 -Ni 5 -Nx 253.00 -Ny 500.00 -Nz 0.00 -Ne
99.998515 -Nl RTR -Nw --- -Ma 0 -Md fffffff -Ms 2 -Mt 800 -Is 2.255 -Id
```

```
-1.255 -It AODV -Il 48 -If 0 -Ii 0 -Iv 29 -P aodv -Pt 0x2 -Ph 2 -Pb 1 -Pd 3
-Pds 0 -Ps 0 -Pss 4 -Pc REQUEST
s -t 2.002819090 -Hs 5 -Hd 2 -Ni 5 -Nx 253.00 -Ny 500.00 -Nz 0.00 -Ne
99.998515 -Nl RTR -Nw --- -Ma 0 -Md 0 -Ms 0 -Mt 0 -Is 5.255 -Id 0.255 -It
AODV -Il 44 -If 0 -Ii 0 -Iv 30 -P aodv -Pt 0x4 -Ph 1 -Pd 3 -Pds -1 -Pl
10.000000 -Pc REPLY
r -t 2.002819335 -Hs 3 -Hd -2 -Ni 3 -Nx 261.00 -Ny 578.00 -Nz 0.00 -Ne
99.998515 -Nl RTR -Nw --- -Ma 0 -Md ffffffff -Ms 2 -Mt 800 -Is 2.255 -Id
-1.255 -It AODV -Il 48 -If 0 -Ii 0 -Iv 29 -P aodv -Pt 0x2 -Ph 2 -Pb 1 -Pd 3
-Pds 0 -Ps 0 -Pss 4 -Pc REQUEST
s -t 2.002819335 -Hs 3 -Hd 2 -Ni 3 -Nx 261.00 -Ny 578.00 -Nz 0.00 -Ne
99.998515 -Nl RTR -Nw --- -Ma 0 -Md 0 -Ms 0 -Mt 0 -Is 3.255 -Id 0.255 -It
AODV -Il 44 -If 0 -Ii 0 -Iv 30 -P aodv -Pt 0x4 -Ph 1 -Pd 3 -Pds 4 -Pl
10.000000 -Pc REPLY
s -t 2.002843829 -Hs 4 -Hd -2 -Ni 4 -Nx 380.00 -Ny 395.00 -Nz 0.00 -Ne
99.998515 -Nl MAC -Nw --- -Ma 0 -Md ffffffff -Ms 4 -Mt 800 -Is 4.255 -Id
-1.255 -It AODV -Il 106 -If 0 -Ii 0 -Iv 29 -P aodv -Pt 0x2 -Ph 2 -Pb 1 -Pd 3
-Pds 0 -Ps 0 -Pss 4 -Pc REQUEST
N -t 2.002844 -n 2 -e 99.998489
N -t 2.002844 -n 1 -e 99.998485
N -t 2.002844 -n 5 -e 99.998485
N -t 2.002844 -n 0 -e 99.998489
N -t 2.002845 -n 3 -e 99.998485
r -t 2.003692138 -Hs 2 -Hd -2 -Ni 2 -Nx 307.00 -Ny 338.00 -Nz 0.00 -Ne
99.998489 -Nl MAC -Nw --- -Ma 0 -Md ffffffff -Ms 4 -Mt 800 -Is 4.255 -Id
-1.255 -It AODV -Il 48 -If 0 -Ii 0 -Iv 29 -P aodv -Pt 0x2 -Ph 2 -Pb 1 -Pd 3
-Pds 0 -Ps 0 -Pss 4 -Pc REQUEST
```

# APPENDIX C

# Trace File Field Types

Field 0: event type

s: send r: receive d: drop f: forward

Filed 1: General tag

–t: time

Field 2: Next hop info

–Hs: id for this node
–Hd: id for next hop towards the destination

Field 3: Node property type tag

–Ni: node id
–Nx –Ny –Nz: node's x/y/z coordinate
–Ne: node energy level
–Nl: trace level, such as AGT, RTR, MAC
–Nw: reason for the event

Field 4: packet info at MAC level

–Ma: duration
–Md: dest's ethernet address
–Ms: src's ethernet address
–Mt: ethernet type

Field 5: Packet information at IP level

–Is: source address. Source port number
–Id: dest address.dest port number
–It: packet type
–Il: packet size
–If: flow id
–Ii: unique id
–Iv: ttl value

Field 6: Packet info at "Application level" which consists of the type of application like ARP, TCP, CBR, the type of ad-hoc routing protocol like DSDV, DSR, AODV etc. The field consists of a leading −P and the list of tags for different applications.

For values of the fields for AODV and CBR are described below;

For AODV:

−Pt : Control message type
−Ph: Hop-count
−Pb: Broadcast-id
−Pd: Destination
−Pds: Dest Seqno
−Ps: Source
−Pss: Source Seqno
−Pl: Lifetime
−Pc: Pkt Type, REPLY/ERROR

For CBR:

−Pn: This denotes the application of "CBR"
−Pi: sequence number
−Pf: how many times this pkt was forwarded
−Po: optimal number of forwards

Printed in the United States
By Bookmasters